*Simplified Estimating
for Builders
and Engineers*

Joseph E. Helton
Texas State Technical Institute

Simplified Estimating

for Builders

and Engineers

PRENTICE-HALL, INC., Englewood Cliffs, New Jersey 07632

Library of Congress Cataloging in Publication Data

HELTON, JOSEPH E., *(date)*
    Simplified estimating for builders and engineers.

    Includes index.
    1. Building—Estimates. I. Title.
TH435.H37 1985   692´.5   84-16085
ISBN 0-13-810144-2

Editorial/production supervision and interior design: Tom Aloisi
Cover design: Jayne Conte
Manufacturing buyer: Tony Caruso
Illustrations and artwork: Terry Conroy
Top right cover photo: Courtesy of Wheeling Corrugating Company

ISBN 0-13-810144-2 01

Prentice-Hall International, Inc., *London*
Prentice-Hall of Australia Pty. Limited, *Sydney*
Editora Prentice-Hall do Brasil, Ltda., *Rio de Janeiro*
Prentice-Hall Canada Inc., *Toronto*
Prentice-Hall of India Private Limited, *New Delhi*
Prentice-Hall of Japan, Inc., *Tokyo*
Prentice-Hall of Southeast Asia Pte. Ltd., *Singapore*
Whitehall Books Limited, *Wellington, New Zealand*

Contents

Preface   xi

1   GENERAL REQUIREMENTS   1

Objectives   1
The Estimating Process   1
The MASTERFORMAT   8
Stretch-Out-Length Concept   9
Work Exercise 1: The Estimating Process   9
Work Exercise 2: Stretch-Out-Length Concept   11

2   SITE WORK AND EXCAVATION   13

Objectives   13
General   13
Subsurface Exploration   14
Site Clearing and Grading   14
Topsoil, Earthwork, and Excavation   17
Swellage and Shrinkage   18
Utility and Drainage Trenches   18
Backfill   20
Roads, Parking, and Walks   21
Fences and Landscaping   22
Work Exercise 3: Excavation   22
Work Exercise 4: Sewer Trench Excavation and Backfill 24

3   CONCRETE   25

Objectives   25
General   25
Types of Cement   26
Concrete Take-offs   26
Footings and Foundation Walls   31
Floor Slabs   35
Beams and Girders   38
Columns   41
Stairs   42
Construction and Control Joints   43
Work Exercise 5: Concrete Stairs Take-off   44
Work Exercise 6: Stretch-Out-Length   44
Work Exercise 7: Detailed Formwork Take-off   47
Work Exercise 8: Formwork, Reinforcement, and Concrete   48
Work Exercise 9: Concrete and Reinforcement   50

4   MASONRY   52

Objectives   52
General   52
Bricks   54
Concrete Blocks   62
Reinforcement   65
Mortar   66
Stone   67
Work Exercise 10: Concrete and Reinforcement, Bricks and Mortar   67
Work Exercise 11: Concrete and Reinforcement, Bricks and Mortar   69
Work Exercise 12: Modular and Standard Bricks, Mortar, and Wall Ties   71
Work Exercise 13: Concrete Blocks by Types and Mortar   73
Work Exercise 14: Modular Bricks and Concrete Blocks   75
Work Exercise 15: Stone and Mortar   77

5   METALS   79

Objectives   79
General   79
Structural Steel   80
Metal Joists   82
Metal Decking   83

Miscellaneous Steel   85
Work Exercise 16: Structural Steel   87
Work Exercise 17: Door and Window Lintels   87

6   WOOD AND PLASTICS   89

Objectives   89
General   89
Rough Carpentry   90
Finish Carpentry   107
Work Exercise 18: Underpinning and Floor System   110
Work Exercise 19: Floor System   112
Work Exercise 20: Foundation and Floor System   113
Work Exercise 21: Plates and Studs   115
Work Exercise 22: Roof Framing   117
Work Exercise 23: Roof Framing   118
Work Exercise 24: Exterior Finish   120
Work Exercise 25: Exterior Finish   123

7   THERMAL AND MOISTURE PROTECTION   125

Objectives   125
General   125
Waterproofing   126
Dampproofing   127
Insulation   127
Roofing   129
Flashing and Sheetmetal Work   131
Caulking   131
Work Exercise 26: Insulation, Building Felt, and Asphalt Shingles   131
Work Exercise 27:   Ceiling Insulation, Building Felt,
      and Asphalt Shingles   133

8   DOORS AND WINDOWS   135

Objectives   135
General   135
Basic Door Types   136
Basic Window Types   141
Work Exercise 28: Doors and Windows   141

9   **FINISHES** **145**

Objectives   145
General   145
Gypsum Drywall   147
Painting   148
Carpet and Pad   150
Resilient Flooring   152
Ceramic Wall and Floor Tile   152
Acoustical Tile   152
Work Exercise 29: Gypsum Drywall, Paint, Carpet, Resilient Flooring,
    and Ceiling Tile   155

10   **MECHANICAL**   **156**

Objectives   156
General   156
Heating and Air Conditioning   157
Plumbing   158
Work Exercise 30: Heating, Ventilation, and Air Conditioning   158
Work Exercise 31: Plumbing   161

11   **ELECTRICAL**   **163**

Objectives   163
General   163
Service-Entrance Equipment   165
Branch Circuits   165
Convenience Outlets and Switches   166
Special-Purpose Items   166
Estimating Electrical Costs   167
Work Exercise 32: Electrical   167

12   **OVERHEAD AND PROFIT**   **170**

Objectives   170
General   170
Overhead   170
Profit   172
Work Exercise 33: Overhead and Profit   173
Work Exercise 34: Final Bid   175

APPENDIX

A: The Stretch-Out-Length Concept   176
B: Formula for Determining Weight of Reinforcing Steel   180
C: Roof Factor Formulas for Common Rafters, and Hip and Valley Rafters   181
D: MASTERFORMAT: Master List of Section Titles and Numbers   185

Index   193

Preface

Regarding the learning process, Samuel Johnson is reported to have said that it is more important to remind than to lecture. It is with this concept in mind that this book has been written. The purpose of this book is to present and demonstrate applications and techniques related to construction estimating.

Although estimating has been defined as a skill involving knowledge and current data, it is much more than that. Estimating also requires judgment based on experience. Therefore, it is impossible to become a skilled estimator through study alone. However, it is also impossible to become a skilled estimator without a basic comprehension of the technical aspects involved. This book's material and method of presentation are directed toward the technical aspects required to make quantity takeoffs in the preparation of an estimate.

The first nine of the twelve chapters of this book follow the numbering sequence of divisions of the MASTERFORMAT of the Construction Specifications Institute. For example, Chapter 3 of the book covers concrete, which is Division 3 of the MASTERFORMAT; Chapter 4 covers masonry, which is Division 4; and so on. Each chapter is followed by work exercises which allow the reader to apply knowledge gained in that particular chapter.

At the beginning of each chapter, objectives are stated. There is also an overall objective of the book, and that is "to prepare the reader to become a skilled estimator capable of making accurate, speedy, and complete quantity take-offs in the preparation of an estimate."

For the long hours of productive work spent on the illustrations and artwork in this book, I want to express my appreciation to Terry Conroy of the Texas State Technical Institute.

For their useful comments, I also wish to express my appreciation to the manuscript reviewers: Professor Robert J. Bradley of the Delaware Technical and Community College; Professor Daniel S. Turner of the University of Alabama; and to Professor John R. Warner of the Springfield Technical Community College.

Joseph E. Helton, P.E.

1

General Requirements

OBJECTIVES

Upon completion of this chapter, the student will be able to:

- List five elements involved in the preparation of the direct costs in a construction estimate.
- Prepare and submit a proposal form.
- Recognize the importance of working drawings and specifications in the preparation of an estimate.
- Recognize the advantages and use of the MASTERFORMAT.

THE ESTIMATING PROCESS

Although the preparation of a construction estimate may involve a variety of approaches and varying degrees of detail, there are five basic elements involved in the estimating process. These elements are: (1) working drawings and specifications, (2) subcontractors' bids, (3) quantity take-offs, (4) checklists, and (5) a summary cost estimate. These five elements involved in the preparation of direct costs are shown in their relationships to one another in Figure 1-1. But before the estimator can arrive at a final bid, factors other than direct costs must also be considered. These factors include financing, land costs, overhead, and profit. After all direct and indirect costs are considered, the contractor submits a bid on the proposal form furnished by the owner. On some occasions, contractors may prepare their own proposal forms. A sample proposal form is shown in Figure 1-2.

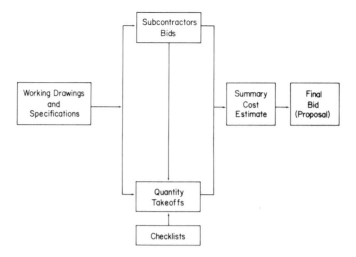

FIGURE 1-1   The Estimating Process

In the process of making a take-off, the estimator should follow the same sequence in which the construction actually takes place on the job. For example, the estimator may begin with general requirements, and then cover site preparation, concrete, and other items.

A review of the five elements involved in the preparation of an estimate will be helpful in understanding the estimating process.

Working Drawings and Specifications

The working drawings and the specifications are the main sources of information in the preparation of a cost estimate for a project. In getting to know what is in the plans and specifications, the estimator may take a few hours, several days, weeks, or even longer, depending on the estimator's past experience, and the complexity and size of the job being considered. The plans and specifications should present a complete description of the structure to be built, details concerning the construction site, the materials to be used, and the construction methods to be employed. Construction drawings usually include site plans, elevations, floor plans, foundation plans, structural framing plans, and details and sections.

Subcontractors' Bids

Before preparing an estimate, the general contractor must decide which construction activities are to be done by subcontractors. A contractor should contact three to six subcontractors for each phase of the work that the contractor's personnel will not do. These subcontractors will be asked to submit competitive bids on their work. One effective and simple way to do this is by sending subcontractors a "bid request" postcard, which identifies the job, the date bids are due, and requests that the subcontractor submit a bid by telephone or letter. Subcontractors usually furnish labor, mate-

PROPOSAL

_____ 19___

Dear Sir:
    The undersigned proposes to furnish all materials and perform all labor necessary to complete the following:

All of the above work to be completed in a substantial and workmanlike manner for the sum of_____
_____ _____($_____) Dollars
    Payments to be made each_____as the work progresses to the value of_____(___%)
per cent of all work completed. The entire amount of contract to be paid within_____days after completion.
    Any alteration or deviation from the above specifications involving extra cost of material or labor will only be executed upon written orders for same, and will become an extra charge over the sum mentioned in this contract. All agreements must be made in writing.
    The Contractor agrees to carry Workmen's Compensation and Public Liability Insurance, also to pay all Sales Taxes, Old Age Benefit and Unemployment Compensation Taxes upon the material and labor furnished under this contract, as required by the United States Government and the State in which this work is performed.
                            Respectfully submitted,

                            Contractor

                            By
ACCEPTANCE

    You are hereby authorized to furnish all materials and labor required to complete the work mentioned in the above proposal, for which the undersigned agrees to pay the amount mentioned in said proposal, and according to the terms thereof.

Date_____19____

PRACTICAL Form 147
MFD. IN U.S.A.

FIGURE 1-2 Proposal Form. (Courtesy of the Frank R. Walker Company)

rials, and equipment required to complete their phases of the work. However, there are occasions when the general contractor may want the subcontractor to bid on labor only. Figure 1-3 shows a sample subcontract agreement form.

Quantity Take-offs

The quantity take-off involves computing the amounts of materials and equipment required to complete a project. To do this, it is necessary to work with an itemized list of the activities involved in building that particular project. These checklists are discussed below.

Although there are a number of ways for a take-off to proceed, the construction sequence is usually the most logical. For example, after the site preparation and excavation activities have been estimated, concrete could be taken next. In the concrete section, the sequence could be pier footings, foundation piers, wall footings, foundation walls, ground slabs, steps, columns, beams, girders, supported slabs, roof fill, floor finishes, rubbing, and curbs. The same procedure could be followed in the other sections as well. Since the estimate for the cost of a project is directly related to the quantity take-off, the validity of the final bid depends, to a large degree, on the accuracy of the individual quantity take-offs. Therefore, it is extremely important that the working drawings and specifications be studied and understood before the quantity take-offs begin.

Checklists

Checklists should be made to fit the job being considered. They will vary in length, detail, and scope depending on the type and complexity of the project or structure. For example, the checklist for a single-family residence will be relatively simple compared to the checklist for a hospital. The purpose of a checklist is to remind the estimator to include every significant item performed in the construction process. A sample checklist is shown for a residence in Figure 1-4.

Summary Cost Estimate

To arrive at a summary cost estimate, it is necessary to combine bids accepted from the various subcontractors with the estimated costs obtained from the quantity take-off times the appropriate unit costs. Also included in the summary is the cost of land, financing, overhead, and profit. Two of these, overhead and profit, will be covered later.

Since realistic unit costs are essential in the preparation of a valid final bid, they should be selected with care. They should be based on the best available information as to how much a unit of work will cost. They should include labor, material, and equipment when applicable. Average cost data are available in estimating reference books such as Means' *Building Construction Cost Data*, Frank R. Walker Company's *The Building Estimator's Reference Book*, and others. However, many estimators prefer to use unit cost figures that are based on their own experience and developed for their own areas of activity. A sample summary cost estimate form is shown in Figure 1-5.

Sub-Contract Agreement

THIS AGREEMENT, made this _____ day of _____ A.D. 19_____,
by and between _____ hereinafter called the
Contractor, and _____ hereinafter called
the Sub-contractor.

For the consideration hereinafter named, the said Sub-contractor covenants and agrees with said Contractor, as follows:

FIRST. The Sub-contractor agrees to furnish all material and perform all work necessary to complete the _____

_____

_____

_____

_____

_____

_____

for the above named structure, according to the plans and specifications (details thereof to be furnished as needed) of
_____ Architect, and to the full satisfaction of said Architect.

SECOND. The Sub-contractor agrees to promptly begin said work as soon as notified by said Contractor, and to complete
the work as follows: _____

_____

_____

_____

_____

_____

THIRD. The Sub-contractor shall take out and pay for Workmen's Compensation and Public Liability Insurance, also
Property Damage and all other necessary insurance, as required by the Owner, Contractor or by the State in which this work
is performed.

FOURTH. The Sub-contractor shall pay all Sales Taxes, Old Age Benefit and Unemployment Compensation Taxes upon the
material and labor furnished under this contract, as required by the United States Government and the State in which this
work is performed.

FIFTH. No extra work or changes under this contract will be recognized or paid for, unless agreed to in writing before
the work is done or the changes made.

SIXTH. This contract shall not be assigned by the Sub-contractor without first obtaining permission in writing from
the Contractor.

IN CONSIDERATION WHEREOF, the said Contractor agrees that he will pay to the said Sub-contractor, in _____
_____ payments, the sum of _____

_____ Dollars
for said materials and work, said amount to be paid as follows: _____ per cent (_____%) of all labor
and material which has been placed in position by said Sub-contractor, to be paid on or about the _____
of the following month, except the final payment, which the said Contractor shall pay to the said Sub-contractor within
_____ days after the Sub-contractor shall have completed his work to the full satisfaction of the said
Architect or Owner.

The Contractor and the Sub-contractor for themselves, their successors, executors, administrators and assigns, hereby agree
to the full performance of the covenants of this agreement.

IN WITNESS WHEREOF, they have executed this agreement the day and date written above.

Witness:

_____

_____

Sub-Contractor
By
Contractor
By

PRACTICAL FORM 133
MFD IN U.S.A.

FIGURE 1-3 Subcontract Agreement. (Courtesy of the Frank R. Walker Company)

Division 1:	*General Requirements*
	Building permit, temporary facilities, insurance, cleanup, closeout
Division 2:	*Site Work*
	Clearing site, site grading, piers and footings, underslab fill, backfilling and grading, loading, trucking
Division 3:	*Concrete*
	Forms, reinforcing steel, concrete: piers and footings, slabs, flatwork, finishing
Division 4:	*Masonry*
	Face and common brick, firebrick, mortar, sand, and cement, cleaning and pointing, wall ties
Division 5:	*Metals*
	Pipe columns, lintels, gable louvers, soffit and roof vents, brackets, hangars, fireplace damper and liner
Division 6:	*Wood and Plastics*
	Rough carpentry: (1) floor systems: sills, girders, headers, subflooring, ledgers, screeds; (2) walls: plates, studs, headers, gables, bracing, blocking, firestops, sheathing, posts, columns, nailers; (3) roof systems: ceiling joists, girders, trusses, headers, common rafters, hip and valley rafters, ridges, collar beams, bracing, sheathing, lookouts, purlins and fascia backing; (4) stairways: treads, risers, stringers, carriage boards, disappearing stairs; (5) furring: plates, studs, strips and lath; (6) rough hardware: nails, anchors, hangers, bolts
	Finish carpentry: posts and beams, siding and paneling, frieze boards, fascia boards, soffits, drywall, flooring, base and shoe molding, ceiling and crown molding, stair and balcony railings, nosings, treads, risers, cabinets and shelving, closet rods, built-in items, hardware, nails, bolts
Division 7:	*Thermal and Moisture Protection*
	Floor and perimeter insulation, wall insulation, ceiling insulation, waterproofing and termite protection, roofing felt, shingles and roofing tiles, fasteners, flashing and sheet metal, caulking and sealants
Division 8:	*Doors and Windows*
	Exterior and interior doors, door frames and trim, sidelights, screen doors, windows and screens, door and window casing, sliding doors and track
Division 9:	*Finishes*
	Sealers and stains, primers, exterior paint, interior paint, wall coverings, taping, floating, and texturing, floor coverings, tile floor and wainscot, counter surfaces and splashes, finishing wood floors
Division 12:	*Furnishings*
	Fabrics, window treatment, furniture and accessories, rugs and mats, interior plants and planters
Division 15:	*Mechanical*
	Plumbing rough-in, heating and air conditioning rough-in, finish plumbing, finish heating and air conditioning
Division 16:	*Electrical*
	Electrical rough-in, electrical finishing

Note: The number divisions of this checklist correspond to the division of construction activities as standardized by the Construction Specifications Institute (CSI); see Appendix D.

FIGURE 1-4   Sample Checklist for a Residence

	SUMMARY COST ESTIMATE				

Project _____ Estimate No. _____
Location _____ Sheet No. _____
Architect/Engineer _____ Date _____
Summary by _____

CSI Div. No.	Description	Estimated Material Cost	Estimated Labor Cost	Subcontracts	Total
1	General Requirements				
2	Site Work				
3	Concrete				
4	Masonry				
5	Metals				
6	Rough Carpentry				
	Finish Carpentry				
7	Thermal and Moisture Protection				
8	Doors				
	Windows				
9	Finishes				
10	Specialties				
11	Equipment				
12	Furnishings				
13	Special Construction				
14	Conveying Systems				
15	Heating, Ventilating, and Air Conditioning				
	Plumbing				
16	Electrical				
	Total Direct Cost				
	Overhead				
	Profit				
	Cost Per Square Foot _____		TOTAL BID		

FIGURE 1-5 Summary Cost Estimate Form

THE MASTERFORMAT

One of the most impressive efforts to standardize the organization of construction project documents has been made by the Construction Specifications Institute (CSI) using the organizational method as currently published in their MASTERFORMAT. A copy of the MASTERFORMAT may be found in Appendix D.

Under the MASTERFORMAT, the U.S. industry CSI and Canadian formats were merged and published as a joint document in 1972. Under this system, project manuals are divided into four parts: (1) bidding requirements, (2) contract forms, (3) conditions of the contract, and (4) specifications. The last part, specifications, is broken down into 16 divisions, which include the following items:

- *Division 1—General Requirements:* A summary of the work schedules and reports, samples and shop drawings, temporary facilities, progress and payment, cleaning up, alternatives, and project closeout
- *Division 2—Sitework:* Clearing of the site, earthwork, soil poisoning, pile foundations, caissons, drainage, utilities, pavements and walks, site improvements, landscaping, railroad, and marine work
- *Division 3—Concrete:* Formwork, joints, reinforcement, cast-in-place concrete, precast concrete, and cementitious decks
- *Division 4—Masonry:* Mortar, unit masonry, stone, and masonry restoration
- *Division 5—Metals:* Structural metal, steel joists, decking, light-gauge framing, ornamental metal, and miscellaneous metal
- *Division 6—Wood and Plastics:* Rough and finish carpentry, heavy timber construction, laminated structural wood, wood treatment, and fabricated structural plastics
- *Division 7—Thermal and Moisture Protection:* Waterproofing, dampproofing, insulation, shingles and roofing tiles, preformed roofing and siding, membrane roofing, flashing and sheet metal, roof accessories, caulking and sealants
- *Division 8—Doors and Windows:* Metal doors and frames, wood and plastic doors, metal windows, wood and plastic windows, hardware and specialties, glazing, curtain walls, and storefronts
- *Division 9—Finishes:* Lath and plaster, gypsum drywall, tile, terrazzo, veneer stone, acoustical treatment, wood flooring, resilient flooring, carpeting, painting, and wall covering
- *Division 10—Specialties:* Chalkboards, chutes, compartments, disappearing stairs, dock facilities, firefighting devices, fireplaces, partitions, lockers, storage shelving, and others
- *Division 11—Equipment:* Items for banks, commercial buildings, darkrooms, churches, educational buildings, laboratories, laundries, libraries, medical, mortuary, musical, prison, and others
- *Division 12—Furnishings:* Artwork, blinds and shades, cabinets and fixtures, drapery and curtains, furniture, and seating

- *Division 13—Special Construction:* Clean rooms, incinerators, greenhouses, insulated rooms, swimming pools, chimneys, vaults, prefabricated structures, nuclear reactors, and others
- *Division 14—Conveying Systems:* Elevators, dumbwaiters, hoists and cranes, lifts, material handling systems, moving stairs and walks, and pneumatic tube systems
- *Division 15—Mechanical:* General provisions, materials and methods, insulation, water supply and treatment, wastewater disposal and treatment, plumbing, fire protection, power or heat generation, refrigeration, controls and instrumentation
- *Division 16—Electrical:* General provisions, materials and methods, power generation, distribution, fixtures, communication, lighting, controls and instrumentation

THE STRETCH-OUT-LENGTH CONCEPT

In the process of making take-offs of lengths, areas, and volumes, the stretch-out-length concept may be used to good advantage by an estimator. This concept can save time and effort in computing concrete volumes, reinforcement lengths, masonry units, and many other items.

The stretch-out length, SOL, is the length of the centerline of any strip of t thickness which bounds the perimeter of a building foundation. The formula for determining the stretch-out length is

$$SOL = P_o - 4t$$

The terms of this formula are defined as

$$SOL = \text{stretch-out length}$$
$$P_o = \text{length of the outside perimeter}$$
$$t = \text{thickness or width of given strip}$$

Regardless of the number of corners or offsets, this formula will give the exact length of the centerline of a given strip of t thickness as long as all corners and offsets are formed by 90-degree turns. For a more detailed explanation of the stretch-out-length concept, refer to Appendix A.

WORK EXERCISE 1: GENERAL REQUIREMENTS (CSI DIVISION 1)

The Estimating Process

Objective: After reviewing Chapter 1, complete the statements below by selecting the best answer for each.

1. In preparing an estimate, the main source of information comes from:

    a. The subcontractors' bids

    b. The quantity take-offs

    c. The working drawings and specifications

    d. The checklists

2. The estimating sequence should begin with which of the following activities?

    a. Electrical work

    b. Site preparation

    c. Carpentry work

    d. Concrete work

3. Which of the following is not an element of the estimating process?

    a. The subcontractors' bids

    b. The quantity take-offs

    c. The contract

    d. The summary cost estimate

4. The most efficient way to get subcontractors to bid on a job is to:

    a. Advertise in the newspaper

    b. Rely on Dodge Reports

    c. Word of mouth

    d. Use bid request postcards

5. Which of the following is not involved in taking off quantities from a set of working drawings?

    a. Volumes

    b. Unit costs

    c. Areas

    d. Pounds

6. The estimator's checklists will vary in length, detail, and scope depending on:

    a. The estimator's experience

    b. The amount of time allowed to complete the project

    c. The type and size of the project

    d. The architect's opinion

7. When preparing a summary cost estimate, the contractor will need all but one of the following:

    a. Subcontractors' bids

    b. Unit costs

    c. Working drawings and specifications

    d. Change orders

8. When preparing the final bid, the contractor will need all but one of the following:

    a. The summary cost estimate

    b. Overhead and profit

c. A building permit

d. The cost of financing

9. The Construction Specifications Institute MASTERFORMAT is arranged to include construction activities in all but one of the following:

a. Financing

b. Excavation and backfill

c. Heating, ventilation, and air conditioning

d. Masonry work

10. In the estimating process, the proposal submitted by a contractor will follow:

a. The summary cost estimate

b. The subcontractors' bids

c. The final bid

d. The checklists

WORK EXERCISE 2: GENERAL REQUIREMENTS (CSI DIVISION 1)

Stretch-Out-Length Concept

Objective: Using the plan view (Figure 1-6) and assuming a height of 8 ft for the basement wall, determine the following and select the correct answer for each.

1. The thickness of the wall is:

a. 6 in.

b. 8 in.

c. 10 in.

d. 12 in.

2. The outside perimeter, P_o, of the wall is:

a. 179.00 ft

b. 194.42 ft

c. 209.83 ft

d. 215.08 ft

3. The stretch-out length, SOL, of the basement wall is:

a. 179.00 ft

b. 205.83 ft

c. 209.83 ft

d. 214.33 ft

4. The area of the top of the basement wall is:

a. 179.00 square feet (SF)

b. 205.83 SF

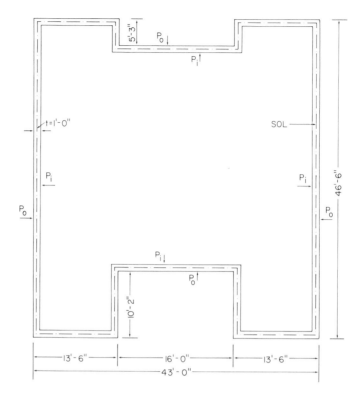

FIGURE 1-6   Plan View of Basement Wall

    c. 209.83 SF

    d. None of the above

5. The volume of concrete required for the basement wall is:

    a. 53.04 cubic yards (CY)

    b. 60.99 CY

    c. 62.17 CY

    d. None of the above

2

Site Work and Excavation

OBJECTIVES

Upon completion of this chapter, the student will be able to:

- Compute the volume of earthwork for site grading on a sample project.
- Compute the number of cubic yards of excavation for a sample trench.
- Compute the volume of earthwork for top soil, bulk excavation, and footings for a sample project.

GENERAL

Since site work involves, among other things, the excavation of materials below the surface, unknown factors involving these materials make this part of estimating difficult. For this reason, the estimator should become familiar with the job site before attempting to estimate excavation quantities. The estimator will have to determine how the actual job site conditions conform with the plot plan. Special attention will have to be given to the type of soil, the amount of visible rock, water conditions, and other factors. Although it may be relatively simple to compute the volumes of earth or rock to be excavated, the costs will depend on factors other than volumes alone. For example, unit costs of excavation will vary considerably between different types of soil, types of rock, moisture conditions, swellage factors, and so on. It will also be necessary to know how the excavation is to be done, and whether equipment may or may not be used.

SUBSURFACE EXPLORATION

In addition to surface inspection, many projects require information on the nature of the soil at various depths below the surface. Borings and test holes are often dug to give data on subsoil conditions. This information is often shown on the working drawings to assist the foundation contractor in the preparation of a bid. Costs will vary for taking soil samples. They may be taken with hand augers or drilling equipment using thin-walled tubes, drilling bits, and wash water.

SITE CLEARING AND GRADING

All items to be removed from the site should be listed under a lump-sum price, or they must be measured. Included in this category of items are trees, shacks and buildings, walls, curbs, fences, and other objects.

In the preparation of a building site, some grading is usually required (Figure 2-1). The site plan is used by the estimator to compute the amount of cut and fill involved. One way to determine the amount of cut or fill is called the borrow pit method. This method requires grid lines to be drawn lightly over the site plan, or alternatively, a transparent plastic overlay is placed over the site plan to avoid marking up the drawing. These grids are drawn to scale, and their intersections are plotted for elevation. These elevations at the intersections of the grid lines are determined by interpolating between the contour lines on the site plan. This procedure is best dem-

FIGURE 2-1   Site Grading

onstrated by an example. Assume that the solid black contour lines shown in Figure 2-2 represent the existing surface. To simplify the example, assume that the finish grade is to be 460 ft. To determine the amount of cut or fill required to reach a finish grade of 460 ft, the procedure is as follows:

1. Draw grid lines vertically and horizontally to a convenient scale. For this example, we will use a 20 ft × 20 ft grid. Across the top, or horizontally, letter the grid intersections a through g. Down the left side, or vertically, number the grid intersections 0 to 4.

2. Using the given contour lines, determine the elevation at each grid intersection. This is done by interpolating between two contour lines. For example, determine the elevation of point a-0 by scaling the perpendicular distance from contour line 462 to a-0. That distance is about 9 ft, and the distance to contour line 463 is about 20 ft. Therefore, the elevation of point a-0 will be 9/20 × 1 plus 462, or 0.45 + 462, or 462.45. In work of this nature, the results are usually rounded off to the nearest one-tenth of a foot, so our answer will be 462.5 ft. Following this procedure, we may obtain the results listed in column (2) of Figure 2-3.

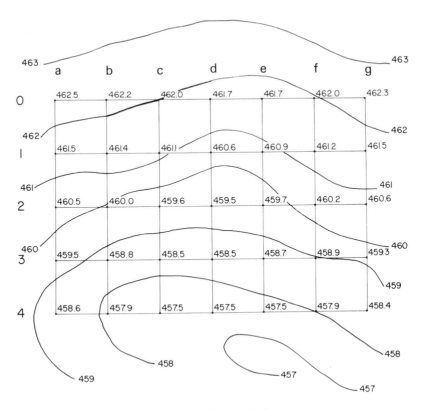

FIGURE 2-2   Site Plan with Contours

(1) Point	(2) Elevation	(3) Cut	(4) Fill	(5) X_n	(6) Cut	(7) Fill
a-0	462.5	2.5		1	2.5	
b-0	462.2	2.2		2	4.4	
c-0	462.0	2.0		2	4.0	
d-0	461.7	1.7		2	3.4	
e-0	461.7	1.7		2	3.4	
f-0	462.0	2.0		2	4.0	
g-0	462.3	2.3		1	2.3	
a-1	461.5	1.5		2	3.0	
b-1	461.4	1.4		4	5.6	
c-1	461.1	1.1		4	4.4	
d-1	460.6	0.6		4	2.4	
e-1	460.9	0.9		4	3.6	
f-1	461.2	1.2		4	4.8	
g-1	461.5	1.5		2	3.0	
a-2	460.5	0.5		2	1.0	
b-2	460.0	0	0	4	0	
c-2	459.6		0.4	4		1.6
d-2	459.5		0.5	4		2.0
e-2	459.7		0.3	4		1.2
f-2	460.2	0.2		4	0.8	
g-2	460.6	0.6		2	1.2	
a-3	459.5		0.5	2		1.0
b-3	458.8		1.2	4		4.8
c-3	458.5		1.5	4		6.0
d-3	458.5		1.5	4		6.0
e-3	458.7		1.3	4		5.2
f-3	458.9		1.1	4		4.4
g-3	459.3		0.7	2		1.4
a-4	458.6		1.4	1		1.4
b-4	457.9		2.1	2		4.2
c-4	457.5		2.5	2		5.0
d-4	457.5		2.5	2		5.0
e-4	457.5		2.5	2		5.0
f-4	457.9		2.1	2		4.2
g-4	458.4		1.6	1		1.6
					53.8	60.0

$$\text{Volume of fill} = \frac{60.0 - 53.8}{4} \times \frac{20 \times 20}{27} = 23.0 \text{ CY (Fill)}$$

FIGURE 2-3   Example of "Borrow Pit" or "Checkerboard" Method for Figuring Volume of Cut or Fill (finish grade: 460.00 ft)

3. Figure the cut or fill by determining the amount the grid intersection elevations are above or below the desired grade of 460 ft. Refer to columns (3) and (4) of Figure 2-3.

4. In column (5) of Figure 2-3, list the number of corners common to each grid

intersection. For example, looking at the site plan of Figure 2-3, note that point a-0 has one corner, b-0 has two corners, c-0 has two, and so on. Also note that some points have four corners, such as b-1, c-1, d-1, and so on.

5. The next step is to multiply the cut-and-fill values in columns (3) and (4) by the values in column (5). In other words, the amount of cut or fill times the number of corners will give the product that goes into column (6) or (7). That is, the cut quantities in column (3) times the values in column (5) will give the values for column (6). In a similar manner, the fill values in column (4) times the values in column (5) will give the values for column (7).

6. Next, total columns (6) and (7), and then take the difference between them. The larger number determines whether you have cut or fill in excess.

7. The algebraic difference obtained in step 6 may now be multiplied by the area of one grid (20 ft × 20 ft), divided by 4, and finally, divided by 27, resulting in the total volume, in cubic yards, of cut or fill required.

TOPSOIL, EARTHWORK, AND EXCAVATION

After the site has been cleared, it is sometimes necessary to strip and store the loam or topsoil covering the area to be graded. In those situations, the volumes of cut or fill estimated for the site grading must be adjusted to compensate for the topsoil thickness that is to be removed. The general practice is to strip the topsoil from all areas of the site to be occupied by buildings, walks, driveways, and other construction features. The topsoil removed is stockpiled for later use.

As mentioned previously, it is relatively easy to compute the number of cubic yards of material to be excavated for a project, but it is much more difficult to determine the cost of excavating the material. The cost will vary according to the type of soil, water encountered, pumping required, bracing of banks, length of haul, and disposal of the excavated material.

On the majority of projects, bulk excavation is accomplished by the use of power-operated equipment, including power shovels, cranes with dragline buckets or clamshells, bulldozers, tractor excavators, scrapers, and trenching machines. In shallow excavations, where the excavated material may be spread not more than 50 ft, bulldozers or tractor excavators may be used.

Bulk or mass excavation for basements and other large areas below grade is determined after the site-clearing operation is estimated. In figuring the volume of the bulk excavation, the depth to the underside of the gravel under the basement slab is measured. Allowance must be made for topsoil removal if it is called for in the specifications. The top grade will be averaged from the grades shown on the plot or site plan. The excavation line around the building will have to be set back to allow for formwork and the sloping of the cut. The sloping of the cut will vary with the soil to be excavated, the depth of the cut, the presence of water, and so on. For soils that have been undisturbed and where water is not present, the slopes used for banks may be 1:1 for sand and gravel, 1:2 for ordinary clay, and 1:3 or 1:4 for stiff clay. These ratios are for the horizontal distance to the vertical distance. In some cases, it may not be possible to slope the banks of an excavation, and then it becomes necessary to sheet and brace the banks to prevent cave-ins.

SWELLAGE AND SHRINKAGE

In computing the cubic volume of an excavation or backfill, provision should be made for swellage and shrinkage. Once earth and rock are disturbed, they begin to swell and increase in volume. When loose material is placed as a backfill material and compacted, it is compressed into a smaller volume, and shrinkage occurs. The swellage and shrinkage varies for different materials. Figure 2-4 gives the range of the swellage factors for four materials.

Material	Swellage Factor
Sand and gravel	1.10–1.18
Loam	1.15–1.25
Dense clay	1.20–1.35
Solid rock	1.40–1.60

FIGURE 2-4   Swellage Factors

To compute the actual amount of earthwork to be handled, the volume of undisturbed soil should be multiplied by the swellage factor for that particular soil or rock. For example, given a loam sample that has a swellage factor of 1.20, the actual volume to be hauled is 1.20 times the undisturbed volume of the soil. If an undisturbed volume of loam soil measures 80 ft × 100 ft and has a depth of 8 ft, then its volume, in cubic yards, would be:

$$\text{Volume} = \frac{80 \times 100 \times 8}{27} = 2370 \text{ CY}$$

But the actual volume of loam soil to be handled would be 1.20 times that amount, or:

$$\text{Volume} = 1.20 \times 2370 = 2844 \text{ CY}$$

The 474-CY increase in volume is due to the swellage of the loam soil during the process of excavation.

UTILITY AND DRAINAGE TRENCHES

In addition to bulk or mass excavation, there are a number of other items that usually require hand excavation or the use of special equipment. The cost per cubic yard for such excavation is more expensive than for bulk excavation. Included in this work are utility trenches (Figure 2-5), footings, drainage trenches, pits, and other special items. Since there is a wide range between unit costs for machine work and handwork, the estimator should list these items separately, using the unit cost figures most appropriate for the excavation to be done. The unit cost of excavation using a trenching machine will vary with the size of the job, width and depth of the trenches, type of soil, and the relative difficulty of getting the equipment into position. In determining the volume of soil to be hauled away, swellage factors should be used.

Drainage Trenches

In site preparation, various types of drainage systems are required to remove subsurface water. Some of the excavation for a drainage system may be handled by the contractor, and some of the excavation may be handled by subcontractors. Since drainage trenches, such as sewer trenches, must have a slope, the computation of the depth of these trenches is based on an average depth.

Subsoil drains are usually vitrified clay pipe set with its lowest point at the bottom of the foundation footings. Although these drains must have a minimum slope of 6 in. per 100 linear feet (LF), a greater slope is normally recommended. The pipe joints for subsoil drains are laid with a $\frac{1}{4}$-in. space between them. The joints are then covered with felt paper or some other suitable material to prevent soil from getting into the pipes.

Calculation of Excavation Volume

To determine the volume of excavation for a sloping trench, refer to the specifications for the slope or pitch for that trench. In Figure 2-6, the slope of a drainage trench is given as $\frac{1}{4}$ in. per linear foot. This means that for every 1 ft of horizontal distance the pipe runs, it drops $\frac{1}{4}$ in. in elevation. Therefore, for a line that runs 100 ft

FIGURE 2-5 Excavating for Electric Cable

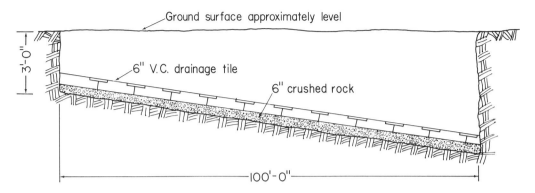

FIGURE 2-6   Drainage Trench Excavation and Backfill

horizontally, the total drop or fall is $\frac{1}{4} \times 100$, or 25 in. To compute the volume of soil to be removed in digging the trench, convert this 25 in. to feet, 2.08 ft, and take the average depth by dividing by 2, giving 1.04 ft for the average depth. To this average depth, add the depth of the trench at the start of the run, or $3.0 + 1.04$. Multiply this sum, 4.04 ft, by the width and length of the trench, and divide by 27 to convert to cubic yards. The result is the undisturbed volume of the excavation. Application of a swellage factor will give the total volume of excavation to be handled. In figuring the total volume of excavation for the drainage trench shown in Figure 2-6, the following information is given. The length is 100 ft, the trench width is 2.0 ft, the starting depth is 3.0 ft, the pitch or slope of the trench is $\frac{1}{4}$ in. per linear foot, and the swellage factor for a dense clay is 1.30. The computation is as follows:

$$\text{Volume (without swellage)} = \left(\frac{2.08}{2} + 3.0\right) \times \frac{2.0 \times 100}{27} = 29.9 \text{ CY}$$

$$\text{Volume (with swellage)} = 29.9 \times 1.30 = 38.9 \text{ CY}$$

BACKFILL

In the process of computing the volume of backfill material, it is necessary to allow for the volume of sand or gravel that is placed as a cushion for slabs, vitrified clay pipe, and so on. This volume must be deducted from the volume of the soil to be used for backfill. Some specifications require the interior backfills to be made only with sand or gravel. The exterior backfill is usually allowed to be excavated material. The selection of machinery and equipment to do backfilling will depend on the type of soil, weather conditions, and how far the backfill material has to be moved.

One of the easiest ways to compute the volume of backfill material is to determine the volume of the constructed foundation within the excavation, and then subtract this volume from the mass or bulk excavation volume, including trench and pit excavations.

Using the drainage trench example shown in Figure 2-6, the backfill required may be computed by determining (1) the volume of the 6-in.-diameter pipe, (2) the volume of the 6-in. layer of crushed rock, (3) taking the sum of these two volumes from the computed excavation volume, and (4) applying a shrinkage factor for the particular soil type.

The shrinkage factor for a material is not applied in the same manner as the swellage factor. To determine the swellage of a material, multiply its bank volume by the swellage factor for that material. To determine the shrinkage of a material, divide its bank volume by 2.00 minus the shrinkage factor for that material. Here is an example:

Assume that the shrinkage factor for a dense clay material is the same as that for its swellage factor, or 1.30. Calculations are as follows:

1. Volume of 6-in.-diameter pipe $= \pi r^2 \times L = 3.1416 \times (0.25)^2 \times 100 = 19.6$ CF.
2. Volume of 6 in. of crushed rock $= 0.5 \times 2.0 \times 100 = 100.0$ CF.

The total of the two volumes equals 119.6 CF, or 4.4 CY.

3. Volume of backfill required $= \dfrac{(29.9 - 4.4)}{(2.00 - 1.30)} = 36.4$ CY.

ROADS, PARKING, AND WALKS

Although the asphalt paving on a project is usually subcontracted, the estimator often makes an estimate in order to check the subcontractor's bid. The estimator will consider subgrade preparation, subdrains, soil stabilization, subbase course, base courses, prime coats, and the asphalt paving needed. Since all of these items are not required on every project, the estimator must review the plans and specifications to determine which are required. In making the estimate, the surface area to be covered is figured in square feet or square yards, and the thickness of each course is noted. In figuring the base course and the asphalt paving, the tonnage required is determined because the ton is the unit in which these items are usually bought.

The unit costs for concrete curbs and gutters is based on units of 100 linear feet for particular cross sections. This unit length multiplied by the unit cost for the particular size of curb and gutter will give the cost for that item. A typical concrete curb and gutter cross section is shown in Figure 2-7.

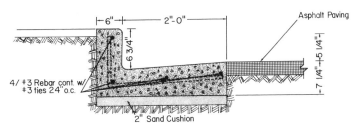

FIGURE 2-7 Typical Concrete Curb and Gutter

FENCES AND LANDSCAPING

Estimating costs for fences is based on a unit cost of so many dollars per linear foot of fence. Landscaping is usually subcontracted, but for those jobs where it is necessary to figure this item, take-offs should include the area of seeding and fertilizing; the area of sodding; the species, sizes, and number of trees and shrubs; and a minimum maintenance period.

WORK EXERCISE 3: SITE WORK (CSI DIVISION 2)

Excavation

Objective: Using Figure 2-8, estimate the amount of excavation, in cubic yards, required for the topsoil, basement, and footings for a building foundation.

Procedure

Study section A-A of Figure 2-8. Note that it shows the layout of a basement wall 12 in. thick and 7 ft 6 in. high. The basement wall sets on a footing which is 1 ft thick and 2 ft wide.

1. Determine how many cubic yards of topsoil must be removed to a depth of 8 in. to the dashed line, which measures 88 ft × 115 ft. In making this computation, it is necessary to include a swellage factor. Assume that the soil is loam with a swellage factor of 1.20. To determine the number of cubic yards of topsoil to be removed, multiply the length, in feet, by the width, in feet, by the depth, in feet, and then divide that product by 27 to convert to cubic yards. The final answer is obtained by multiplying that result by 1.20, the swellage factor.

2. Compute the amount of earth to be removed in order to form the basement walls. Note that it will be necessary to extend the area of earth to be removed 3 ft beyond the exterior walls. This is necessary in order to have room to install the wall forms. The basement excavation is to be dug to a depth of 6 ft 10 in. below the existing grade after the topsoil is removed.

   To calculate the basement excavation, it is convenient to divide the basement area into rectangles. The depth will be the same for all three rectangles. Compute the volumes of earth within each of the three rectangles, and add them together. After converting this volume to cubic yards, multiply by the swellage factor, 1.20, to get the total volume of earth to be removed from the basement.

3. Determine how many cubic yards of earth must be removed for the footing for this foundation. Assume that the footing will be dug 12 in. deep and that no swellage factor will be used. The reason for this is that the material dug for the footing is a relatively small volume which usually does not have to be hauled from the job site.

   Use the stretch-out-length method to determine the length of the footing. Multiply that value by the width of the footing, in feet, and the depth of the footing, in feet. Dividing that product by 27 will give the number of cubic yards of earth to be removed for the footing. Compute the amount of earth removed for the footing.

Work Exercise 3

23

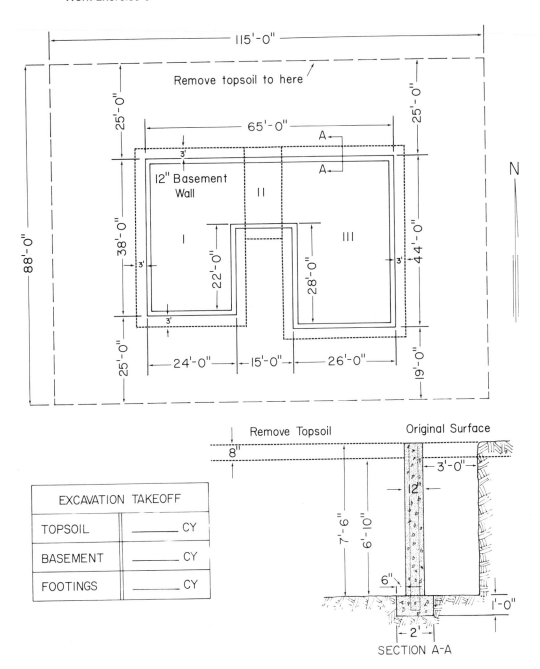

FIGURE 2-8 Excavation

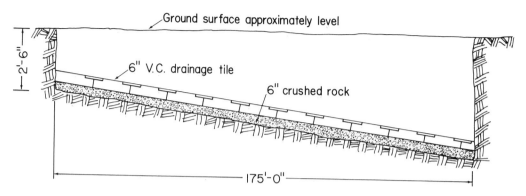

FIGURE 2-9 Sewer Trench Excavation and Backfill

WORK EXERCISE 4: SITE WORK (CSI DIVISION 2)

Sewer Trench Excavation and Backfill

Objective: Using the sketch in Figure 2-9 and the information below, make a take-off of the excavation and backfill required in order to install a sewer line.

Given

Starting depth, 2 ft 6 in.
Trench width, 2 ft
Slope or pitch, $\frac{1}{4}$ in. per linear foot
Swellage factor, 1.25
Shrinkage factor, 1.20
Horizontal length, 175 ft

3

Concrete

OBJECTIVES

Upon completion of this chapter, the student will be able to:

- Make a take-off of the formwork required to form footings, foundation walls, floors and slabs, beams and girders, columns, and stairs in a sample reinforced concrete structure.
- Compute the volume of concrete required for each of the items listed above.
- Compute the number of pounds of steel reinforcement required for the items listed above.
- Compute the number of square feet of finishing required for the items listed above.

GENERAL

Freshly mixed concrete is made up of aggregates and a paste which consists of portland cement and water. The aggregates are usually sand and gravel and comprise from 60 to 75% of the total volume of structural concrete. Some desired properties of fresh concrete are consistency, uniformity, and workability. Some desired properties of hardened concrete are durability, strength, watertightness, and resistance to abrasion. A concrete mix that is properly designed will have the required qualities of hardened concrete, strength and durability; it will have workability during placement; and it will be economical.

TYPES OF CEMENT

The properties of concrete may be controlled by careful selection and proportioning of the basic ingredients of the concrete mix. There are five types of portland cement: Type I, which is normal cement used for general purposes; Type II, or modified, used to reduce temperature rise and to increase resistance to sulfate attack; Type III, or high early strength, used when it is necessary to remove forms early, or in cold weather; Type IV, or low-heat, used on large gravity dams and other large concrete masses where the heat of hydration must be kept to a minimum; and Type V, or sulfate-resistant, used in concrete exposed to severe sulfate action encountered in some western states. Of the five types, Type I is the general-purpose cement and the one most often used in construction work.

Since concrete should be durable, strong, watertight, and abrasion resistant, attention must be given to the quality of the portland cement paste. The quality of the paste is determined by the total amount of water mixed with the cement. This is the "water-cement ratio" and is expressed in terms of the ratio of the weight of the water to the weight of the cement.

A large volume of the concrete used in today's construction is ready-mixed and transported to the job site in special trucks. The concrete may be mixed at a central mixing plant or in ready-mix trucks enroute to the job site. At the job site, the concrete is placed in forms directly or dumped into hoppers for distribution by wheelbarrows, buggies, or crane-hoisted buckets.

Although concrete varies depending on the type of aggregates used, 1 cubic foot of concrete usually weighs between 145 and 150 lb. For special use, some lightweight concrete weighs from 70 to 110 lb per cubic foot. There are a number of factors which influence the cost of concrete. For instance, reinforced concrete, which has steel bars or welded wire fabric embedded in it, will cost more than "plain concrete." Other factors, such as strength requirements and admixtures, will also add to the cost of concrete.

CONCRETE TAKE-OFFS

In making a concrete take-off, the estimator usually refers to the structural drawings, where the dimensions for footings, foundation walls, floors and slabs, beams and girders, columns, stairs, and so on, are to be found. Once these measurements are known, the volumes of the various items are easily determined. The primary components of a reinforced concrete building are shown in Figure 3-1. Included in a concrete take-off are formwork, reinforcement, cast-in-place concrete, and finishing.

Formwork

The cost of the formwork as well as the cost of the labor to install and remove it is to be included in the take-off. Concrete formwork is usually figured separately for each

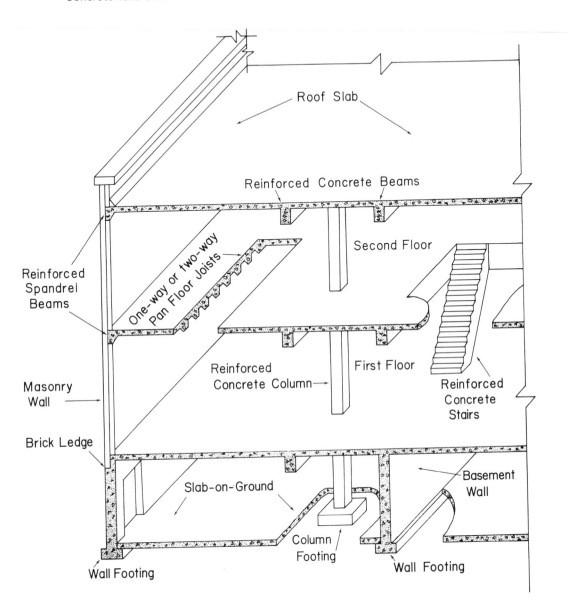

Roof Slab

Reinforced Concrete Beams

Reinforced Spandrel Beams

One-way or two-way Pan Floor Joists

Second Floor

Masonry Wall

Reinforced Concrete Column

First Floor

Reinforced Concrete Stairs

Brick Ledge

Slab-on-Ground

Basement Wall

Wall Footing

Column Footing

Wall Footing

FIGURE 3-1 Components of a Reinforced Concrete Building

section of the concrete work. For example, footings, foundation walls, slabs, beams, columns, and stairs are listed separately. Although formwork for footings is sometimes omitted when the excavation is very hard and the banks will stand without caving in, most specifications require that footing forms be used.

Concrete wall forms are used to hold wet, plastic concrete until it has set or cured sufficiently to allow removal of the forms safely. Where possible, the forms are reused for other parts of the construction. The number of reuses depends on a number of factors, including the material used, type of coatings, and the care taken to prevent damage to the forms during stripping.

Concrete forms should be strong and rigid, well braced, and oiled or coated to provide quick release during stripping. In addition, the use of coatings and overlays on plywood forms increases both the number of reuses and the quality of the concrete finish.

The amount of formwork required for a specific item is based on and directly related to the surface area of forms actually in contact with the concrete. This area is referred to as the square feet of contact area and is given the abbreviation SFCA.

The material and labor costs per square foot of contact area varies with the thickness and height of wall as well as the number of reuses that may be obtained from the forms. On many small jobs, contractors find it convenient and economical to use foundation wall forms made up into panels measuring 1 ft × 8 ft, 2 ft × 8 ft, and 4 ft × 8 ft. The frames for these panels are made with 2 × 4 lumber and sheathed with $\frac{3}{4}$-in. Plyform plywood. The corners are sometimes reinforced with galvanized straps for greater strength and durability.

On larger jobs, a contractor may find it more economical to use various types of prefabricated panel forms such as those manufactured by the Symons Corporation, the Burke Company, the Ceco Corporation, and others. Some of these forms combine plastic-coated plywood with steel frames. They are excellent for good-quality finishes and for a great number of reuses. These forms come in 2-ft widths and heights ranging from 3 to 8 ft.

Reinforcement

The cost of reinforcement and the labor and equipment involved in its handling and placement must be included in the take-off.

Although concrete is strong in compression and shear, it is relatively weak in tension. For most structural uses, concrete is reinforced with steel bars or welded wire fabric. In some cases, concrete can handle low tensile stresses and does not need reinforcement. When concrete is placed without steel reinforcement, it is called "plain concrete."

However, in most commercial construction, steel reinforcement is required in concrete. This reinforcement may be in the form of bars or welded wire fabric, sometimes called "wire mesh." Commonly used for reinforcing slabs on grade and for temperature reinforcement on suspended slabs, welded wire fabric comes in a square pattern of 4 in. × 4 in., 6 in. × 6 in., and so on. It is normally estimated by weight in pounds per 100 SF.

Reinforcing steel bars are estimated by the pound or ton. Therefore, it is necessary to determine the linear footage of steel bars by size. This footage is then con-

Bar Number	Bar Size (in.)	Nominal Diameter (in.)	Area (sq. in.)	Weight (PLF)
3	$\frac{3}{8}$	0.375	0.11	0.376
4	$\frac{1}{2}$	0.500	0.20	0.668
5	$\frac{5}{8}$	0.625	0.31	1.043
6	$\frac{3}{4}$	0.750	0.44	1.502
7	$\frac{7}{8}$	0.875	0.60	2.044
8	1	1.000	0.79	2.670
9	$1\frac{1}{8}$	1.128	1.00	3.400
10	$1\frac{1}{4}$	1.270	1.27	4.303
11	$1\frac{3}{8}$	1.410	1.56	5.313
14	$1\frac{3}{4}$	1.693	2.25	7.650
18	$2\frac{1}{4}$	2.257	4.00	13.600

FIGURE 3-2   Weights of Standard Reinforcing Bars

verted to pounds by using the weight per linear foot for that particular size of steel bar. Figure 3-2 gives the weights per foot for standard reinforcing bars. The number of a steel reinforcing bar is based on the number of eighths of an inch in the diameter of that particular bar.

For example, a No. 3 bar is one that has three eighths in its diameter, or a nominal diameter of $\frac{3}{8}$ in. Similarly, a No. 4 bar has a nominal diameter of $\frac{1}{2}$ in. (four eighths), and a No. 18 bar has a nominal diameter of $2\frac{1}{4}$ in. (18 eighths).

If reference data are unavailable, an easy way to approximate the weight per foot for steel reinforcing bars is to use the formula

$$\text{Weight per foot for a steel bar} = 2.67D^2$$

where D is equal to the diameter of the bar in inches. For an explanation of this formula, refer to Appendix B.

Reinforced concrete structures are generally designed so that the separate parts act as a single unit. Therefore, it is necessary to locate construction joints and provide steel reinforcement through the joints. Since splices in reinforcement cannot be avoided in most situations, it is necessary to provide additional length where joints are required. Most bars may be obtained in lengths up to 60 ft, but transporting these lengths is often a problem and shorter lengths are more common.

Three types of splices are commonly used: lapped, welded, and mechanical. In most cases, lapped splices are the most economical. The length of lap varies with the concrete strength, the yield strength of the steel, and the bar size. The length of the lap is usually given in the specifications or on the working drawings. Many specifications call for a minimum length of splice of 12 in. or 24 bar diameters, whichever is

Bar Number	Minimum Lap (24D)
3	1'-0"
4	1'-0"
5	1'-3"
6	1'-6"
7	1'-9"
8	2'-0"
9	2'-4"
10	2'-7"
11	2'-10"

FIGURE 3-3 Minimum Laps

greater. For bar sizes 3 through 11, the minimum lengths of splices are as shown in Figure 3-3.

Cast-in-Place Concrete

The concrete take-off must include the cost of the concrete for a specific item and the cost of the labor and equipment for mixing and placing it. In making a concrete take-off, the estimator should convert the volumes involved into cubic yards. Since the dimensions of footings, walls, and beams are normally given in feet and inches, the lengths, widths, and depths of these items multiplied together give the volume in cubic feet. To convert to cubic yards, divide cubic feet by 27, the number of cubic feet in 1 CY.

Finishing

This part of the concrete take-off must include the cost of finishing, curing, and protection in cold weather; and it should include material, labor, and equipment. In estimating the cost of concrete finishing, the first step is to determine the type and square footage of the predominant finish. For floors and slabs, the predominant finish will be noted on the floor finish schedule or in the specifications.

Proper curing is a key factor in obtaining quality concrete. One of the best ways to cure concrete is by leaving the forms in place for at least 4 or 5 days. If this is not possible, an efficient curing method is to spray or brush on a curing compound immediately after the surface of the concrete has been troweled. This provides an airtight seal over the surface of the concrete. Other methods of curing involve the use of damp burlap, sand, sawdust, or ponding.

Estimating the cost of curing depends to a large extent on knowing what type of curing will be required. Moisture retaining and watertight covers are estimated by the square foot of surface to be covered.

FOOTINGS AND FOUNDATION WALLS

A cross section of a typical footing and foundation wall is shown in Figure 3-4. Footings are designed to distribute loads transmitted to them from the building above. The principle of spreading loads over a wide area results in an increased load-bearing capacity. By setting the foundation wall on a footing with twice the width, the load-bearing capacity is doubled. The thickness of the footing is dependent on the load-bearing capacity of the material on which it rests. A common practice is to design the footing thickness to be equal to the width of the wall which it supports.

In making take-offs of foundation walls and footings, it is sometimes difficult to determine their total lengths. A method that simplifies this task is the stretch-out-length concept introduced in Chapter 1 and explained in Appendix A.

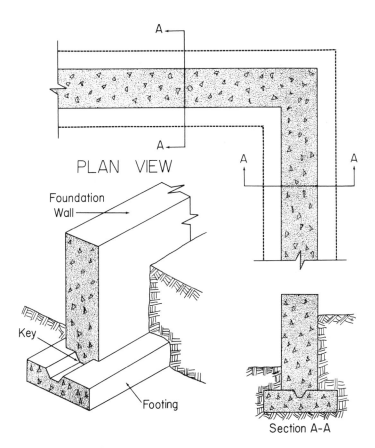

PLAN VIEW

Foundation Wall

Key

Footing

Section A-A

FIGURE 3-4 Typical Footing and Foundation Wall

Formwork

Forms for concrete footings, foundation walls, and retaining walls may be estimated by determining the number of square feet of contact area (SFCA) between the forms and the wet concrete. Since forms are normally used on both sides of footings and walls, the SFCA is determined for one side and then doubled. A formula for finding the SFCA for a footing or wall with forms on both sides is as follows:

$$SFCA = SOL \times height \times 2$$

In figuring the SFCA for a foundation wall with openings, these openings are not normally deducted unless they are greater than 30 SF on one contact surface. To compute the SFCA for a column footing, multiply the total length around the footing by the depth of the footing.

Although many estimators prefer to base the cost of formwork on the square footage of contact area, it is sometimes necessary to have a more detailed breakdown of the individual items involved in building forms. For instance, it may be desirable to figure the number of board feet of lumber required for bracing, studs, and walers. In addition, the number of form ties and square footage of Plyform may be figured. In cases where this type of detail is required, these items may be estimated as follows.

Studs and Bracing.   In wall forms, vertical studs are computed by dividing the stretch-out length of the wall by the stud spacing. This result gives the number of studs that have lengths the same as the height of the wall. To find the number of studs for both sides, multiply the result by 2. Individual bracing is computed on the basis of its spacing, size, and use. For bracing that is placed at a 45-degree angle to the wall, the length of each brace is approximately the height of the wall multiplied by the square root of 2, or 1.4.

Walers.   Walers are doubled 2×4 members running horizontally or vertically and spaced as close as necessary to resist the lateral pressures on the wall form. The number of walers on one side of the form is determined by dividing the height of the wall by the spacing of the walers when they are run horizontally. When the walers are run vertically, the number of walers is determined by dividing the stretch-out length of the wall by the waler spacing. In both cases, the results are doubled where forms are being used on both sides.

The length of walers running horizontally is equal to the stretch-out length of the wall. The length of walers running vertically is equal to the height of the wall.

Form Ties.   To estimate the number of form ties, determine the horizontal and vertical spacing of the ties. This information is usually given in the formwork details. In those cases where spacings are not given, the estimator or contractor may use spacings based on their experience. An example in estimating form ties is shown using the layout in Figure 3-5. In this figure, the horizontal spacing of the form ties is 32 in.

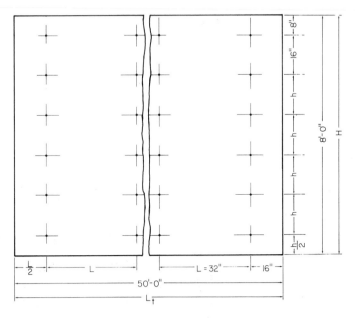

FIGURE 3-5   Spacing of Form Ties

on centers, and the vertical spacing is 16 in. on centers. The spacings will vary for different situations and for different lateral pressures. Regardless of the spacings, the approximate number of ties for a given contact area is found by using the following formula:

$$\text{Number of form ties } = \frac{H}{h} \times \frac{L_t}{L}$$

where H = height of the wall
$\quad L_t$ = total length of the wall
$\quad h$ = vertical spacing of the ties
$\quad L$ = horizontal spacing of the ties

Using the example in Figure 3-5, the values are:

$$\begin{aligned}
H &= 8 \text{ ft} \\
L_t &= 50 \text{ ft} \\
h &= 16 \text{ in.} = 1.33 \text{ ft} \\
L &= 32 \text{ in.} = 2.67 \text{ ft}
\end{aligned}$$

Substituting these values into the formula, the number of ties required is:

$$\text{Number of ties (approximate)} = \frac{8}{1.33} \times \frac{50}{2.67} = 113 \text{ ties}$$

Plywood. Normally, the square footage of plywood required for a foundation wall and many other items will be the same as the square footage of the SFCA. This value, divided by 32, gives the number of pieces of plywood 4 ft × 8 ft. The material cost per piece or per square foot will depend on the thickness, type, and grade of plywood specified.

Reinforcement

To estimate the cost of reinforcement in footings and foundation walls, the estimator must refer to the working drawings. In the details and cross sections, the number, sizes, and placement of steel reinforcing will be given. The estimator will compute the total linear feet of each size bar. This total length will include laps for joints. The total length is then multiplied by the weight per linear foot for that particular size bar. In determining the weight per foot for a steel bar, the estimator may refer to tables in various references like that in Figure 3-2, or the following formula may be used to approximate the weight per foot of bars:

$$W = 2.67D^2$$

where W is the weight of steel bar in pounds per linear foot, and D is the diameter of the steel bar in inches. For example, to find the weight per foot of a No. 6 steel bar, the formula is

$$W = 2.67(0.75)^2 = 1.502 \text{ lb/ft}$$

For information on the derivation of this formula, see Appendix B.

The stretch-out length (Appendix A) may also be helpful in computing the linear feet of steel bars in a footing or foundation wall. The SOL for a particular footing or wall, multiplied by the number of horizontal bars, plus a lap factor, will give the total number of linear feet for that bar. This result may then be converted to the total number of pounds of steel required for that particular bar size.

Concrete

The cost of cast-in-place concrete is related directly to the volume of concrete required, job conditions, methods of placement, and whether the concrete is job-mixed or ready-mixed. Because of the wide variability of these factors, concrete costs are subject to a wide range. Other factors include function and size of the building, types of slab construction, building codes, and so on. However, the first responsibility of the estimator is to come up with an accurate figure for the volume of concrete required.

Finishing

The cost of finishing, curing, and protection of footings and wall foundations will depend on the finishing specifications for a particular job. The cost of finishing is usually based on the square footage of area involved.

FIGURE 3-6 On-Ground Slab

FLOOR SLABS

Concrete on-ground slabs (Figure 3-6) vary in thickness from 3 to 8 in. They are generally laid over a well-tamped layer of gravel, crushed stone, or cinders. Supported slabs, like slabs-on-ground, are measured over their total surface area and multiplied by their thickness to get the total volume of concrete required. However, not all slabs are flat or plate types, and the computation of their volumes is a little more difficult. Some types of designs of reinforced concrete buildings have a concrete joist system which uses steel or plastic pans to facilitate forming. The working drawings and specifications will indicate the type of slab system and give the dimensions necessary to determine concrete volumes.

Formwork

In figuring the formwork for slabs, it is necessary to refer to the working drawings and determine the type of slab. There are flat slabs, plate slabs, one-way joist system slabs, two-way joist system slabs, and other types. If the slab is a suspended flat slab as shown in Figure 3-7, the formwork is usually figured as the total area of the underside of the slab. The sides of the slab are figured with the beams or girders that the slab frames into. The formwork for an on-ground slab is figured as the perimeter of the slab, in feet, times the slab thickness, in feet. The result is the number of square feet of contact area, or SFCA for that slab.

When metal pans are used to form one-way (Figure 3-8) or two-way (Figure 3-9) joist systems, the pans are specified to be not less than 14-gauge slip-in removable steel forms when a quick release is required. These adjustable-type pans allow the contractor to remove the pans without disturbing the shoring that supports the

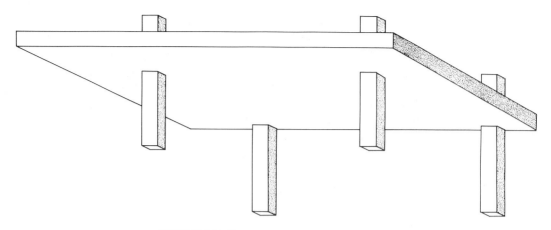

FIGURE 3-7 Suspended Flat Slab

weight of the slab system. The shapes and sizes of these metal pans are given on the working drawings.

Reinforcement

Reinforcing steel in flat and plate-type slabs may be determined by figuring a weight per square foot based on the spacing and size of the steel bars. The information in Figure 3-10 gives the pounds of reinforcing steel per square foot for a wall or slab area. For example, in a slab where No. 3 bars are spaced 18 in. on centers both ways,

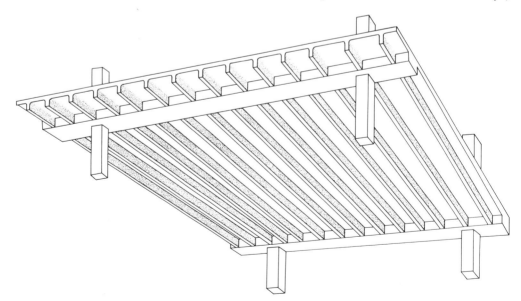

FIGURE 3-8 One-Way Floor Joist System

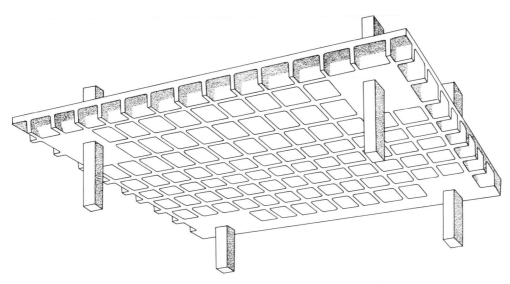

FIGURE 3-9   Two-Way Floor Joist System

the weight of the steel reinforcing is 0.25 lb per square foot (PSF). The total weight of the reinforcing steel in the slab is figured by multiplying this value, 0.25 PSF, times the total number of square feet in the slab.

Cast-in-Place Concrete

In estimating the concrete volume for flat and plate-type slabs, the surface area of the slab times the thickness of the slab, plus a waste factor when required, will give the number of cubic yards.

   In figuring the concrete for one-way and two-way joists systems, many estimators rely on published weights and volumes made available by various manufacturers

| | Spacing, Center to Center, Both Ways (in.) | | | | | | | |
Bar No.	4	6	8	10	12	14	16	18
2	0.50	0.33	0.25	0.20	0.17	0.14	0.13	0.11
3	1.13	0.75	0.56	0.46	0.38	0.32	0.28	0.25
4	2.00	1.34	1.00	0.80	0.67	0.57	0.50	0.45
5	3.13	2.09	1.56	1.25	1.04	0.89	0.78	0.70
6	4.51	3.00	2.25	1.80	1.50	1.29	1.13	1.00
7	6.13	4.09	3.07	2.45	2.04	1.75	1.53	1.36
8	8.01	5.34	4.01	3.20	2.67	2.29	2.00	1.78
9	10.20	6.80	5.10	4.08	3.40	2.91	2.55	2.27
10	12.91	8.61	6.45	5.16	4.30	3.69	3.23	2.87
11	15.94	10.63	7.97	6.38	5.31	4.55	3.98	3.54

FIGURE 3-10   Pounds of Reinforcing Steel per Square Foot of Wall or Slab Area

of steel pans. Figure 3-11 shows one of these tables published by the CECO Corporation. Using this type of information, the estimator may determine the volumes and weights per square foot of floor area for various sizes of steel pans in a one-way joist system. For example, for a 30-in.-wide steel pan, a 3-in.-thick slab, a joist 5 in. wide and 12 in. deep, the concrete volume is 0.422 CF per square foot of floor. Tables are also available for two-way joist systems.

Finishing

Where concrete slabs are used as a base for asphalt, vinyl, rubber, or linoleum flooring, it is imperative that a smooth, level finish be obtained. The cost is directly related to the method or methods used to obtain a particular finish quality. During recent years, great improvements have been made in finishing quality through the use of power-operated finishing machines, and by the use of admixtures in concrete. If finishing must be done by hand, the estimator may allow 70 to 90 SF of floor area per hour per finisher, subject to job conditions.

BEAMS AND GIRDERS

Interior beams are an integral part of the slab which they support and are T-shaped in cross section. Reinforced concrete spandrel beams that span openings in the outer walls of a building must be able to support bricks or concrete blocks that form a building's exterior walls. Intermediate, or interior, beams are also a part of floor slabs, but they are located within the building, as shown in Figure 3-12. Beams sometimes frame into larger structural members called "girders."

Formwork

The formwork for beams and girders is figured as the square footage of the beam sides plus the beam bottom. The basic difference in figuring the SFCA for a spandrel beam as opposed to an intermediate beam is that the outside of the spandrel beam will be wider than its inside. Both sides of an intermediate beam are usually of the same depth.

Reinforcement

Concrete protective covering for steel reinforcement in beams, girders, columns, and slabs should be at least $1\frac{1}{2}$ in. In estimating the linear feet of reinforcing steel in a beam or girder, use the lengths and bends shown on the details of the working drawings. If a bar length is not given, use the approximate length that will include the

FLANGEforms

All sizes shown conform to the Concrete Reinforcing Steel Institute, CRSI, Code of Standard Practice, MSP-1-76.

Intermediate FLANGEforms are furnished in 20″ and 30″ standard widths in 3-ft. lengths as standard. One-and two-foot lengths are available as filler lengths.

Endcaps are used at the ends of rows of Intermediate FLANGEforms at beams, bridging joists and special headers.

Tapered Endforms are used at the ends of joists where extra joist width is required. They are furnished only in 20″ and 30″ standard widths, and in standard 3-ft. lengths only.

Each side of 20″ wide Tapered Endforms tapers 2″ in 3-ft., providing a 4″ increase in joist width. The 30″ wide Tapered Endforms taper 2½″ providing a 5″ increase in joist width.

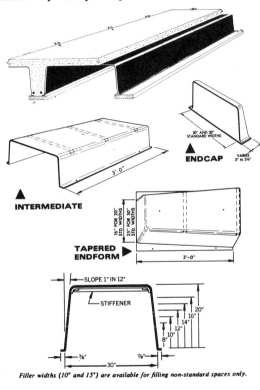

INTERMEDIATE

ENDCAP   VARIES 3″ to 5½″

20″ AND 30″ STANDARD WIDTHS

TAPERED ENDFORM

16″ FOR 20″ STD WIDTHS 25″ FOR 30″ STD WIDTHS

3′-0″

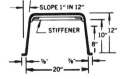

SLOPE 1″ IN 12″

STIFFENER

20″
16″
14″
12″
10″
8″

⅞″   ⅞″

30″

Filler widths (10″ and 15″) are available for filling non-standard spaces only.

SLOPE 1″ IN 12″

STIFFENER

10″ 12″
8″

⅞″   ⅞″

20″

CONCRETE QUANTITIES/30″ WIDTHS*

Depth of Steelform	Width of Joist	Cubic feet of concrete per square foot for various slab thicknesses			Additional concrete for Tapered Endforms. cu. ft. per lin. foot of bearing wall or beam (One side only)
		2½″	3″	4½″	
8″	5″	.317	.359	.484	.143
	6″	.333	.374	.499	.139
	7″	.347	.389	.514	.135
10″	5″	.348	.390	.515	.179
	6″	.367	.409	.534	.174
	7″	.386	.427	.552	.169
12″	5″	.381	.422	.547	.214
	6″	.404	.445	.570	.208
	7″	.425	.467	.592	.203
14″	5″	.415	.456	.581	.250
	6″	.441	.483	.608	.243
	7″	.467	.508	.633	.236
16″	6″	.481	.522	.647	.278
	7″	.509	.551	.676	.270
	8″	.537	.578	.703	.263
20″	6″	.564	.606	.731	.347
	7″	.599	.641	.766	.338
	8″	.633	.675	.800	.329

Apply only for areas over FLANGEforms and joists between them. Bridging joists, special headers, beam tees, etc., not included.

CONCRETE QUANTITIES/20″ WIDTHS*

Depth of Steelform**	Width of Joist	Cubic feet of concrete per square foot for various slab thicknesses			Additional concrete for Tapered Endforms. cu. ft. per lin. foot of bearing wall or beam (One side only)
		2½″	3″	4½″	
8″	4″	.339	.381	.506	.167
	5″	.361	.402	.527	.160
	6″	.380	.422	.547	.154
10″	4″	.377	.419	.544	.208
	5″	.404	.445	.570	.200
	6″	.428	.470	.595	.192
12″	4″	.418	.459	.584	.250
	5″	.449	.491	.616	.240
	6″	.479	.520	.645	.231

Apply only for areas over FLANGEforms and joists between them. Bridging joists, special headers, beam tees, etc., not included.

**14,″ 16″ and 20″ depths are also available. Contact your Ceco Sales Engineer.*

VOIDS CREATED BY VARIOUS SIZE FLANGEFORMS
Shaded areas below indicate standard filler widths

Depth of Steelform	Cubic feet of void created per linear foot (for various widths of FLANGEforms)				**Cubic Feet per Tapered End	
	30″ Width	20″ Width	15″ Width	10″ Width	30″ Width	20″ Width
8″	1.628	1.072	.794	.516	4.465	2.882
10″	2.023	1.329	.982	.634	5.548	3.569
12″	2.414	1.581	1.165	.748	6.617	4.243
14″	2.801	1.829	1.343	.857	7.673	
16″	3.183	2.072	1.516	.961	8.715	
20″	3.933	2.544	1.850	1.155	10.756	

**Total void created by standard 3′-0″ length Tapered Endform.*

FIGURE 3-11   One-Way Joist Construction. (Courtesy of the CECO Corporation.)

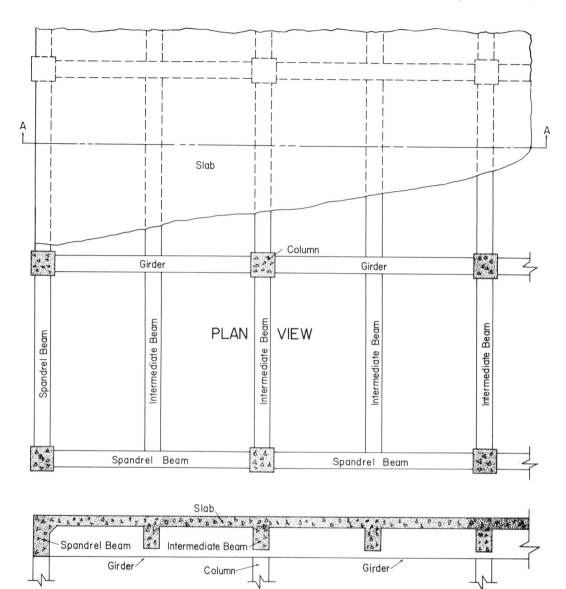

FIGURE 3-12 Beams and Girders in a Building

additional length required for bending the bar. Also include the steel bar required for stirrups on beams and girders. The reinforcing steel required for spandrel beams may be figured by taking the perimeter of the building, adding a factor for laps and bends, and multiplying that result by the number of bars in the beam cross section.

Cast-in-Place Concrete

Although methods vary, many estimators figure the volume of spandrel beams by taking their full depths and measuring the slab to the inside of the spandrel beam. In figuring the depths of intermediate beams, the depths are taken as the distance from the bottom of the beam to the bottom of the slab. Once the depth of the beam is known, its volume is found by multiplying that depth times its width times its length. This result, in cubic feet, is then converted to cubic yards by dividing by 27. A waste factor of 2 to 8%, depending on job conditions, may be added to the computed volume.

Finishing

All exposed concrete surfaces may require some patching or finishing. The amount and degree of work involved will vary depending on the specifications. Rubbed finishes, sandblasting, or bushhammering are a few of the possible finishes that may be specified in the finishing schedule. The items are usually figured by the surface area, or square footage involved.

COLUMNS

Columns may be round, rectangular, L-shaped, or of various irregular cross sections. Irregular shapes are formed by attaching special inserts inside square or rectangular forms. Forms for round columns may be built of wood, but ready-made forms of metal or fiber are commonly used.

Formwork

Square or rectangular columns are usually formed with four panel sides. The column forms may be erected in place, panel by panel, or they may be assembled into a complete box and set in place as a complete unit. The cost of the forms will depend on which method is used, but in either case, the cost will be figured by the square feet of contact area, SFCA. In figuring the SFCA, the true height of the column form is the story height less the slab thickness. This height times the distance around the column sides will give the SFCA.

Reinforcement

Reinforcing for columns may be assembled in place or prefabricated in cages. These cages may be set in place before or after the column forms are in position. For heavy reinforcing bars or large columns, it is common practice to build the reinforcing cage in place. The cost of the reinforcing steel will be based on the total number of pounds of steel required and on the method used in placing the steel.

Cast-in-Place Concrete

The cross-sectional area, in square feet, of a column times its true height, in feet, divided by 27, will give the number of cubic yards of concrete required.

Finishing

Just as for beams and girders, the cost of finishing columns will depend on the finishing requirements called for in the specifications and the degree of finishing quality required. The cost of finishing is based on the number of square feet involved.

STAIRS

Concrete stairs (Figure 3-13) are usually built after the structural frame of the building is completed. The stairs may be precast, but usually they are cast in place.

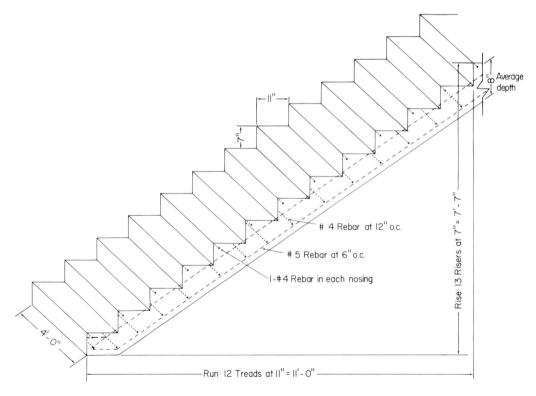

FIGURE 3-13   Concrete Stairs

Formwork

Forms for stairs must be strong, rigid, and tight; and they must be constructed to facilitate their removal without damage to the concrete. Forms for stairs may be lumber and/or plywood, or they may be steel. The square feet of contact area, SFCA, for stair forms may be figured by adding together the areas of the underside, ends, and risers.

Reinforcement

A number of factors, including loading, width, and rise, determine the amount and size of reinforcing steel in a set of stairs. The estimator must rely on the working drawings and specifications in figuring the number of pounds of steel reinforcing required.

Cast-in-Place Concrete

The number of cubic yards of concrete required for a set of stairs may be figured by multiplying the width of the stairs by its length times its average depth. A rule of thumb for determining the average depth is to add 1 in. to the size of the riser. For example, for a set of stairs with 7-in. risers, the average depth may be figured as 8 in. as an approximation. Figure 3-13 shows a set of stairs with risers of 7 in. and treads of 11 in. Using an average depth of 8 in. will give reasonable results in computing the volume of concrete for this set of stairs.

Finishing

The finishing operation on stairs begins at the top and progresses downward. Treads should be floated so that there is a pitch of about $\frac{1}{4}$ in. from back to front to ensure proper drainage. The edger should then be run along the riser board, followed by the first troweling of the tread. The unit cost of finishing a set of stairs is based on the surface area to be finished, the type of finish, and job conditions.

CONSTRUCTION AND CONTROL JOINTS

Construction joints are stopping places in the concrete operation. They are designed to allow for expansion and contraction between different sections of concrete. Although they are usually made with a keyway to provide for load transfer across the joint, dowels may also be used for that purpose. Construction and control joints are estimated on the basis of the number of linear feet of joint. Joint fillers for construction joints may be strips of bituminous-saturated fiber or $\frac{1}{2}$-in. bituminous filler which should be as wide as the slab is thick.

A control joint is a type of joint that encourages cracking along a predetermined straight line or lines spaced certain distances apart. These joints may be formed by tooling or inserting metal or fiber strips immediately after the slab has been placed. Since this operation interferes with the finishing process, a more satisfactory method for forming control joints is to saw them into the concrete after it has stiffened but before it gets too hard. The resulting void is then filled with molten lead or some other filler material to prevent chipping of the edges.

WORK EXERCISE 5: CONCRETE (CSI DIVISION 3)

Concrete Stairs Take-off

Objective: Using Figure 3-13, which shows a set of concrete stairs, make a take-off for the following items:

1. Volume of concrete required (include a 5% waste factor)
2. Pounds of No. 4 reinforcing steel (include an 8% factor for laps)
3. Pounds of No. 5 reinforcing steel (include an 8% factor for laps)

WORK EXERCISE 6: CONCRETE (CSI DIVISION 3)

Stretch-Out Length

Objective: Using the stretch-out-length concept and Figure 3-14, compute the following items and select the correct answer for each.

1. The stretch-out length for the footing and wall is:
   a. 180 ft
   b. 258 ft
   c. 260 ft
   d. None of the above
2. The area of the top of the 6-in. wall that runs around the building is:
   a. 129 SF
   b. 258 SF
   c. 130 SF
   d. None of the above
3. The area of the top of the footing that supports the foundation wall of the building is:
   a. 129 SF

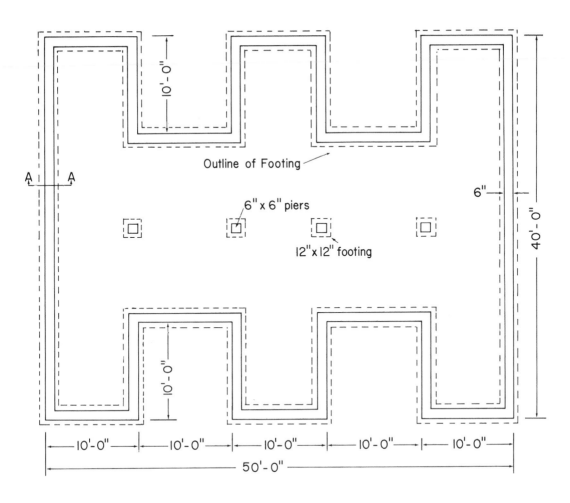

Outline of Footing

6" x 6" piers

12"x 12" footing

10'-0"

6"

40'-0"

A A

10'-0"

10'-0"

10'-0"

10'-0"

10'-0"

10'-0"

50'-0"

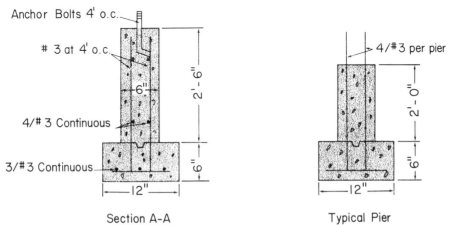

Anchor Bolts 4' o.c.

3 at 4' o.c.

6"

2'-6"

4/# 3 Continuous

3/# 3 Continuous

6"

12"

Section A-A

4/#3 per pier

2'-0"

6"

12"

Typical Pier

FIGURE 3-14 Foundation Plan and Details

   b. 130 SF

   c. 258 SF

   d. None of the above

4. The volume of the foundation wall is:

   a. 322.5 CF

   b. 645 CF

   c. 325 CF

   d. None of the above

5. The volume of the footing is:

   a. 64.5 CF

   b. 65 CF

   c. 129 CF

   d. None of the above

6. The number of cubic yards of concrete required for the foundation wall is:

   a. 11.9 CY

   b. 23.9 CY

   c. 30.0 CY

   d. None of the above

7. The number of cubic yards of concrete required for the footing is:

   a. 2.4 CY

   b. 3.0 CY

   c. 4.8 CY

   d. None of the above

8. The number of anchor bolts required using a waste factor of 5% is:

   a. 68

   b. 75

   c. 88

   d. None of the above

9. The number of pounds of reinforcing steel required, using a lap factor of 5%, for the No. 3 bars placed horizontally in the footing is:

   a. 244 lb

   b. 286 lb

   c. 310 lb

   d. None of the above

10. The number of pounds of reinforcing steel required, using a lap factor of 5%, for the No. 3 bars placed vertically and horizontally in the foundation wall is:

   a. 410 lb

   b. 570 lb

   c. 640 lb

   d. None of the above

WORK EXERCISE 7: CONCRETE (CSI DIVISION 3)

Detailed Formwork Take-off

Objective: Referring to Figure 3-15 and using the information given below, make a detailed formwork take-off of the lumber and nails required to build the formwork for the foundation of a building.

In making this take-off, apply waste factors for the various items as follows:

2×6 stringers, 7%
2×2 stakes, 12%
1×3 spreaders, 16%
Nails, 10%

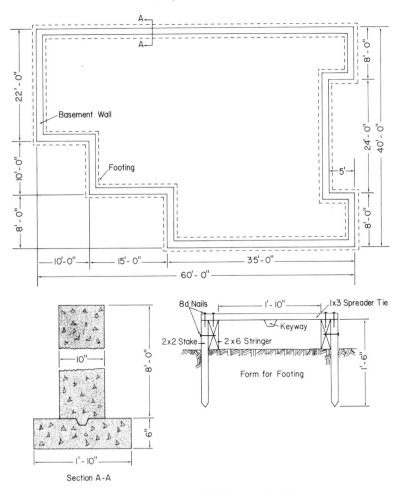

FIGURE 3-15 Foundation Plan and Details

Procedure

Study the plan view of the foundation and the details in Figure 3-15. Determine the size of the footing and what material is to be used in forming.

1. Compute the lumber required for stringers, noting that two 2 × 6 stringers are required. Determine the stretch-out length of the footing and multiply by 2. Apply the waste factor of 7% and divide by 16 to get the number of 2 × 6's which are 16 ft long. Finally, convert this result to board feet. (See Chapter 6 for an explanation of how to convert dimension lumber to board feet.)

2. Compute the lumber required for spreader ties. Determine their length and then determine the number that is needed if they are spaced 4 ft on centers around the length of the footing. The spreaders are made from 1 × 6's which are ripped down the center, and they are 8 ft long. Figure that six spreaders may be obtained from each 1 × 6 that is 8 ft long. Apply a waste factor of 16% and give an answer for the number of 1 × 6 × 8 ft that will be required. Convert the result to board feet.

3. Compute the lumber required for stakes, which are 2 × 2 × 18 in. and spaced 4 ft on centers. Use a waste factor of 12%. Convert the result to board feet.

4. Compute the number of pounds of eight penny (8d) nails required assuming a waste factor of 10%. Also assume that there are approximately 100 8d nails to a pound of nails. Figure 6 nails every 4 ft around the footing length.

WORK EXERCISE 8: CONCRETE (CSI DIVISION 3)

Formwork, Reinforcement, and Concrete

Objective: Given the information in Figures 3-15 and 3-16, make a take-off of the material required to form the basement wall, place the reinforcement steel, and place the concrete in a foundation.

Procedure

Study the figures and note the elements that make up the basement wall forms: plates, walers, tie-rods, bracing, and so on. Also note that these forms are to be prefabricated in three sizes, 1 ft × 8 ft, 2 ft × 8 ft, and 4 ft × 8 ft. In order to simplify the problem, assume that it takes 13 forms of the 1 ft × 8 ft size, 10 forms of the 2 ft × 8 ft size, and 95 forms of the 4 ft × 8 ft size. Using those figures, compute the number of 2 × 4 × 8 ft studs required using a waste factor of 7%. Compute the number of plywood panels required using a waste factor of 5%.

- *Plates:* Two 2 × 4 plates run around the top, and two 2 × 4 plates run around the bottom of the basement wall as shown in Figure 3-15. Use a waste factor of 7%.
- *Walers:* Use 2 × 4 walers as shown in Figure 3-16 with a waste factor of 7%. The 2 × 4's are ordered in 16-ft lengths.
- *Tie-rods:* Every 4 ft around the foundation wall, there are four tie-rods. Use a waste factor of 5%.
- *Anchor bolts:* Figure anchor bolts on a spacing of 4 ft on centers. Add two extra bolts for each corner plus a waste factor of 5%.

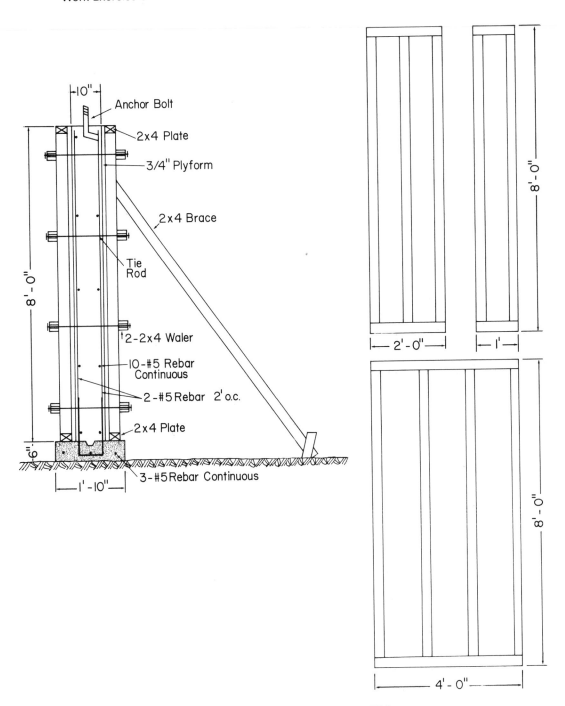

FIGURE 3-16 Forming Basement Wall

- *Bracing:* In figuring the bracing, use 2 × 4's that are 10 ft long and spaced every 4 ft along the length of the wall form. Apply a 10% waste factor.
- *Nails:* Six penny (6d) common nails are used to fasten the plywood to the studs and plates. In nailing the plates to the studs, 16d common nails are used. Use 16d duplex nails for easy removal in fastening the bracing to the form and to the stakes. Figure 15 pounds of 6d common nails for every 1000 SF of plywood required. Figure 20 lb of 16d common nails for every 1000 board feet (BF) of plates required. Figure 10 lb of 16d duplex nails for every 1000 BF of bracing required. For all three types of nails, apply a waste factor of 10%.
- *Reinforcement:* In making a take-off of the reinforcement, note that in Figure 3-15 all the rebar is one size, No. 5. The horizontal steel is continuous as shown, and the vertical steel is spaced 2 ft on centers around the length of the basement wall. Apply a waste factor of 8% to cover laps and waste.
- *Concrete:* In determining the number of cubic yards of concrete required for the footing and the basement wall, use a waste factor of 8%.

Compute the following:

1. Number and board feet of 2 × 4 × 8 ft stud material
2. Number of pieces and square feet of 4 ft × 8 ft Plyform
3. Number and board feet of 2 × 4 × 16 ft plate material
4. Number and board feet of 2 × 4 × 16 ft waler material
5. Number of tie-rods
6. Number of anchor bolts
7. Number and board feet of 2 × 4 × 10 ft bracing material
8. Pounds of 6d common nails
9. Pounds of 16d common nails
10. Pounds of 16d duplex nails
11. Pounds of No. 5 rebar for footings (horizontal only)
12. Pounds of No. 5 rebar for basement wall (horizontal and vertical steel, including that which goes down into the footing)
13. Cubic yards of concrete for the footing
14. Cubic yards of concrete for the basement wall

WORK EXERCISE 9: CONCRETE (CSI DIVISION 3)

Concrete and Reinforcement

Objective: Using Figure 3-17, make a take-off of the concrete, reinforcing steel, and cushion sand required for a building foundation.

In making this take-off, use the following waste factors:

Concrete, 8%
Rebar, 8%
Sand, 15%

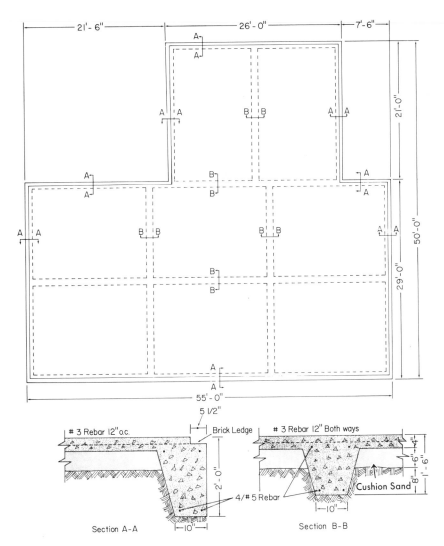

FIGURE 3-17 Foundation Plan and Details

Compute the following:

1. Cubic yards of concrete for exterior-grade beams (A-A) (figure an average width of 12 in.)
2. Cubic yards of concrete for interior beams (B-B) (figure an average width of 14 in.)
3. Cubic yards of concrete for the 4 in. slab
4. Cubic yards of concrete for total foundation (sum of items 1, 2, and 3)
5. Cubic yards of cushion sand
6. Pounds of No. 3 rebar for the slab
7. Pounds of No. 5 rebar for the exterior and interior beams

4

Masonry

OBJECTIVES

Upon completion of this chapter, the student will be able to:

- Compute the number of concrete blocks and bricks required for a small building.
- Determine the amount of reinforcement required for the masonry units in a small building.
- Compute the amount of mortar required to lay masonry units in a small building.

GENERAL

In the process of preparing to make a masonry take-off, the estimator must study the working drawings and specifications and determine where various types of bricks, concrete blocks, stone, and other masonry materials are located in a building or project. In addition to masonry units such as bricks and concrete blocks, other masonry materials include lintels, precast concrete units, coping, trim, and clay tile.

Although masonry take-offs may be difficult because of the large variety of types and sizes of partitions and exterior walls, the basic take-off procedure for each type and size is relatively simple. A masonry take-off usually involves the use of one of three methods. The three methods are (1) the wall area method, (2) the course method, and (3) the volume method.

Wall Area Method

Because of its simplicity and accuracy, the most widely used method is the wall area method. With this method, the estimator either computes, or takes from prepared data as shown in Figure 4-1, the number of blocks or bricks and mortar required to fill the area of 1 SF of a single-wythe wall. (A single-wythe wall is a wall that has a thickness of one masonry unit). The next step is to determine the net wall area that is to receive that particular masonry unit. The net wall area is found by subtracting the area of wall openings from the gross wall area.

Course Method

The course method involves determining the number of courses required for a wall of a given height, and then figuring how many bricks or concrete blocks are in one course. Multiplying these two figures together and subtracting the number of bricks or blocks not needed in openings will give the number of units for that particular wall.

Volume Method

The third method, the volume method, is similar to the wall area method except that the number of bricks or blocks is figured for 1 CF rather than for 1 SF. With this method, it is necessary to know how many masonry units fill 1 CF, and then to figure how many cubic feet are in a given wall.

| | | | CUBIC FEET OF MORTAR | |
Name	Size of Brick (in.)	Number of Bricks per 100 SF	per 100 SF	per 1000 Bricks
		Nonmodular		
Standard	$3\frac{3}{4} \times 2\frac{1}{4} \times 8$	655	5.8	8.8
Nonstandard	$3\frac{3}{4} \times 2\frac{3}{4} \times 8$	551	5.0	9.1
		Modular		
Modular	$4 \times 2\frac{2}{3} \times 8$	675	5.5	8.1
Engineer	$4 \times 3\frac{1}{3} \times 8$	563	4.8	8.6
Norman	$4 \times 2\frac{2}{3} \times 12$	450	5.1	11.2
Jumbo closure	$4 \times 4 \times 8$	450	4.2	9.2
Roman	$4 \times 2 \times 12$	600	6.5	10.6
"SCR brick"	$6 \times 2\frac{2}{3} \times 12$	450	7.9	17.5

Note: Running bond, no waste, and $\frac{3}{8}$ in. of mortar.

FIGURE 4-1 Nonmodular and Modular Bricks and Mortar Required for Single-Wythe Walls

In each of these methods, the estimator must allow for wall openings. In addition, waste factors based on job and handling conditions, should be applied.

Before presenting examples of the three methods, some basic information concerning masonry materials will be reviewed. The materials include bricks, concrete blocks, reinforcement, mortar, and stone.

BRICKS

Bricks are the most widely used masonry unit, and they come in a wide variety of types, textures, sizes, and colors. Figure 4-2 shows a few of the many types of modular and nonmodular bricks used today.

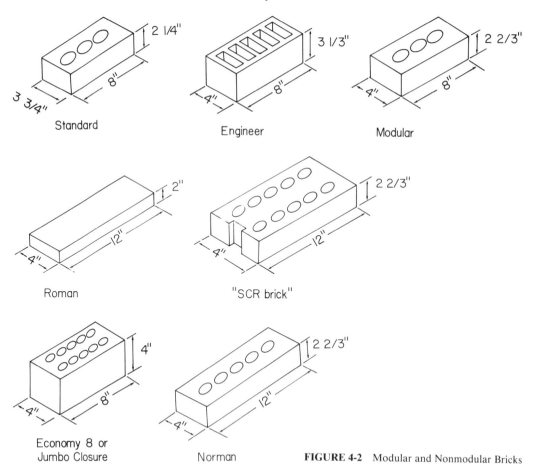

FIGURE 4-2   Modular and Nonmodular Bricks

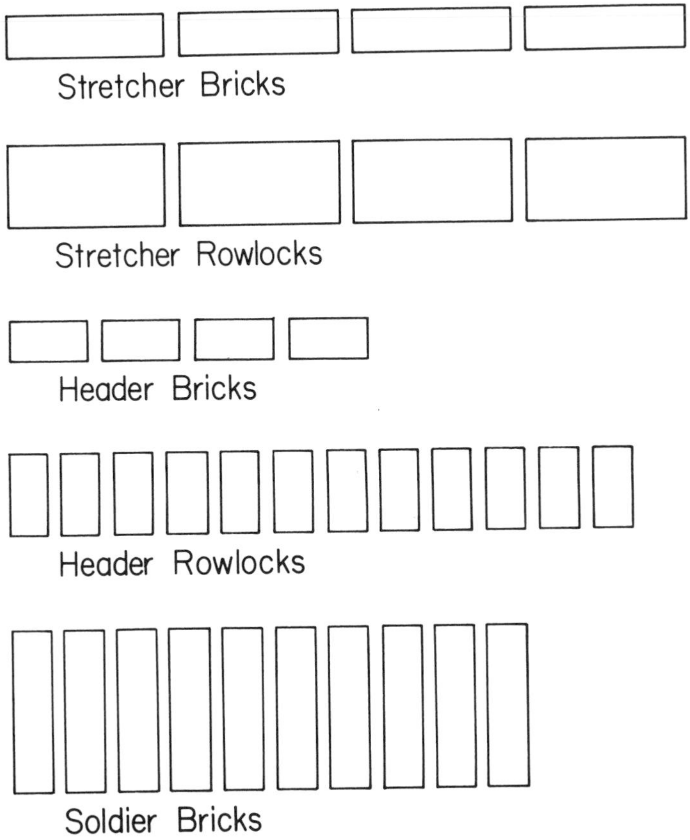

Stretcher Bricks

Stretcher Rowlocks

Header Bricks

Header Rowlocks

Soldier Bricks

FIGURE 4-3 Positioning of Bricks

Not only brick size, but the way bricks are laid will affect the estimator's masonry take-off. Some of the positions in which bricks may be placed are shown in Figure 4-3. Stretcher bricks are laid with their long dimensions showing on the face of the wall. Bricks that are placed vertically or on their ends are called "soldier" bricks. When bricks are laid at right angles to stretcher bricks, they are called "headers." When they are laid on their sides exposing their full widths and lengths, they are called "stretcher rowlocks." When only the widths and thicknesses are exposed, the bricks are in a position called "header rowlocks."

When stretchers and headers are used in combination, the result is a brick bond. A brick bond is simply a pattern to create a pleasing appearance as well as structural strength in a masonry wall. Some bond patterns are shown in Figure 4-4. The masons in the picture in Figure 4-5 are laying a common bond pattern in the wall of an apartment house.

In estimating the number of bricks using the wall area method, the course method, and the volume method, the procedures are as described below.

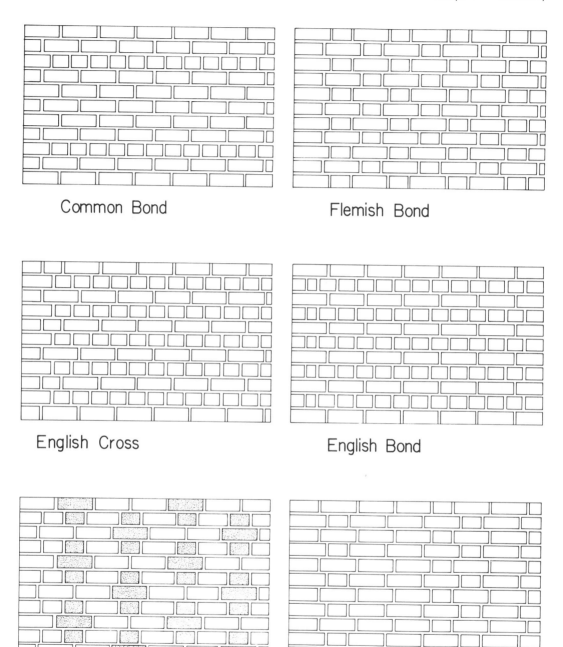

Common Bond

Flemish Bond

English Cross

English Bond

Flemish Cross Bond

Garden Wall Bond

FIGURE 4-4   Some Types of Brick Bonds

FIGURE 4-5 Common Bond Pattern

Estimating Bricks Using the Wall Area Method

The first step is to determine the number of bricks that will lay up in one square foot of the wall area. This may be done, based on experience, reference to tables, or by figuring. If the number of bricks per square foot is to be figured, calculate the area of the brick surface exposed in the face of the wall and add the area of one vertical and one horizontal layer of mortar for that brick. The total area, in square inches, of the brick surface plus the mortar joints is divided into 144, the number of square inches in one square foot. For example, to compute the number of standard bricks with $\frac{3}{8}$-in. mortar joints in a running bond without headers, multiply the height, $2\frac{1}{4}$ in., of a standard brick times its length, 8 in., adding $\frac{3}{8}$ in. for the mortar joints. The resulting height and length are $2\frac{5}{8}$ in., or 2.625 in., and $8\frac{3}{8}$ in., or 8.375 in. This may be clearer after referring to Figure 4-6.

The number of standard bricks in 1 SF may be calculated as follows:

$$\text{Number of standard bricks per SF} = \frac{\text{number of sq. in. in 1 SF}}{\text{face area of brick} + \text{mortar (sq. in.)}}$$

For a standard brick, the number of bricks in 1 SF is equal to

$$\frac{144}{2.625 \times 8.375} = 6.55 \text{ bricks per SF}$$

FIGURE 4-6   Number of Standard Bricks in One Square Foot: 6.55

When standard bricks are laid in a running bond without headers, and when a $\frac{3}{8}$-in. mortar joint is used, 6.55 bricks are required for every square foot of wall surface. Another way of saying the same thing is to say that 655 bricks are required for every 100 SF of wall surface. If a full course of header bricks is used every fifth, sixth, or seventh course, it is necessary to allow for extra brick. In addition, a waste factor is usually included when estimating brick requirements.

The same procedure is used in figuring the number of modular bricks with this important difference. In figuring the standard bricks, it was necessary to add the mortar joints to the brick dimensions. In figuring modular bricks, the given size of the bricks include the mortar joints; therefore, they are not added. Refer to Figure 4-7 and note that 675 modular bricks are required for every 100 SF of wall surface with a running bond without headers.

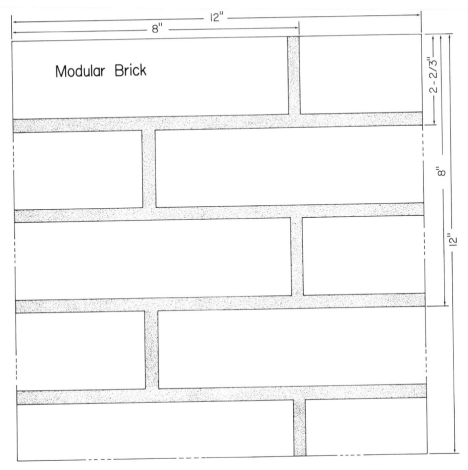

FIGURE 4-7   Number of Modular Bricks in One Square Foot: 6.75

Once the number of bricks per square foot, or per 100 SF, is known, multiply this number times the net wall area, and add an appropriate waste factor. This will give the total number of bricks required for that particular wall surface.

Estimating Bricks Using the Course Method

When using the course method, the first step is to determine the number of bricks in one course. This is done by dividing the length of the wall, in inches, by the length of one brick plus, in the case of standard bricks, the thickness of its end mortar joint, in inches. For example, if the length of a wall is 28 ft 7 in., and its height is 8 ft 9 in., the number of standard bricks in one course would be figured as follows:

$$\frac{\text{Number of standard bricks}}{\text{per course}} = \frac{\text{length of wall (in.)}}{\text{length of brick + mortar joint (in.)}}$$

Assuming a mortar joint of $\frac{3}{8}$ in. thickness:

$$\frac{\text{Number of standard bricks}}{\text{per course}} = \frac{28.58 \times 12}{8.375} = 40.95, \text{ or 41 bricks per course}$$

The next step is to find the number of courses of bricks by dividing the height of the wall by the thickness of one brick plus its mortar joint. This may be expressed as a formula:

$$\text{Number of standard brick courses} = \frac{\text{height of wall (in.)}}{\text{thickness of one brick + mortar joint (in.)}}$$

$$\text{Number of standard brick courses} = \frac{8.75 \times 12}{2.625} = 40 \text{ courses}$$

To find the number of bricks in a wall that is 28 ft 7 in. long and 8 ft 9 in. high without any openings, multiply the number of bricks in one course, 41, by the number of courses, 40:

$$\text{Number of standard bricks} = 41 \times 40 = 1640 \text{ bricks}$$

When openings are present in the wall surface, allowance must be made by deducting the number of bricks that would have been used in those openings. As an alternative to calculating the number of courses for various wall heights, tables similar to Figure 4-8 are available.

Course Number	Standard Brick ($2\frac{1}{4}$ in.)	Modular Brick ($2\frac{2}{3}$ in.)	Concrete Block ($7\frac{5}{8}$ in.)
1	$0'\text{-}2\frac{5}{8}''$	$0'\text{-}2\frac{11}{16}''$	$0'\text{-}8''$
2	$0'\text{-}5\frac{1}{4}''$	$0'\text{-}5\frac{5}{16}''$	$1'\text{-}4''$
3	$0'\text{-}7\frac{7}{8}''$	$0'\text{-}8''$	$2'\text{-}0''$
4	$0'\text{-}10\frac{1}{2}''$	$0'\text{-}10\frac{11}{16}''$	$2'\text{-}8''$
5	$1'\text{-}1\frac{1}{8}''$	$1'\text{-}1\frac{5}{16}''$	$3'\text{-}4''$
6	$1'\text{-}1\frac{3}{4}''$	$1'\text{-}4''$	$4'\text{-}0''$
7	$1'\text{-}6\frac{3}{8}''$	$1'\text{-}6\frac{11}{16}''$	$4'\text{-}8''$
8	$1'\text{-}9''$	$1'\text{-}9\frac{5}{16}''$	$5'\text{-}4''$
9	$1\text{-}11\frac{5}{8}''$	$2'\text{-}0''$	$6'\text{-}0''$
10	$2'\text{-}2\frac{1}{4}''$	$2'\text{-}2\frac{11}{16}''$	$6'\text{-}8''$
11	$2'\text{-}4\frac{7}{8}''$	$2'\text{-}5\frac{5}{16}''$	$7'\text{-}4''$
12	$2'\text{-}7\frac{1}{2}''$	$2'\text{-}8''$	$8'\text{-}0''$

FIGURE 4-8   Heights of Masonry Unit Courses for Brick and Block ($\frac{3}{8}$-in. mortar bed joints)

Course Number	Standard Brick ($2\frac{1}{4}$ in.)	Modular Brick ($2\frac{2}{3}$ in.)	Concrete Block ($7\frac{5}{8}$ in.)
13	$2'\text{-}10\frac{1}{8}''$	$2'\text{-}10\frac{11}{16}''$	$8'\text{-}8''$
14	$3'\text{-}0\frac{3}{4}''$	$3'\text{-}1\frac{5}{16}''$	$9'\text{-}4''$
15	$3'\text{-}3\frac{3}{8}''$	$3'\text{-}4''$	$10'\text{-}0''$
16	$3'\text{-}6''$	$3'\text{-}6\frac{11}{16}''$	$10'\text{-}8''$
17	$3'\text{-}8\frac{5}{8}''$	$3'\text{-}9\frac{5}{16}''$	$11'\text{-}4''$
18	$3'\text{-}11\frac{1}{4}''$	$4'\text{-}0''$	$12'\text{-}0''$
19	$4'\text{-}1\frac{7}{8}''$	$4'\text{-}2\frac{11}{16}''$	$12'\text{-}8''$
20	$4'\text{-}4\frac{1}{2}''$	$4'\text{-}5\frac{5}{16}''$	$13'\text{-}4''$
21	$4'\text{-}7\frac{1}{8}''$	$4'\text{-}8''$	$14'\text{-}0''$
22	$4'\text{-}9\frac{3}{4}''$	$4'\text{-}10\frac{11}{16}''$	$14'\text{-}8''$
23	$5'\text{-}0\frac{3}{8}''$	$5'\text{-}1\frac{5}{16}''$	$15'\text{-}4''$
24	$5'\text{-}3''$	$5'\text{-}4''$	$16'\text{-}0''$
25	$5'\text{-}5\frac{5}{8}''$	$5'\text{-}6\frac{11}{16}''$	$16'\text{-}8''$
26	$5'\text{-}8\frac{1}{4}''$	$5'\text{-}9\frac{5}{16}''$	$17'\text{-}4''$
27	$5'\text{-}10\frac{7}{8}''$	$6'\text{-}0''$	$18'\text{-}0''$
28	$6'\text{-}1\frac{1}{2}''$	$6'\text{-}2\frac{11}{16}''$	$18'\text{-}8''$
29	$6'\text{-}4\frac{1}{8}''$	$6'\text{-}5\frac{5}{16}''$	$19'\text{-}4''$
30	$6'\text{-}6\frac{3}{4}''$	$6'\text{-}8''$	$20'\text{-}0''$
31	$6'\text{-}9\frac{3}{8}''$	$6'\text{-}10\frac{11}{16}''$	$20'\text{-}8''$
32	$7'\text{-}0''$	$7'\text{-}1\frac{5}{16}''$	$21'\text{-}4''$
33	$7'\text{-}2\frac{5}{8}''$	$7'\text{-}4''$	$22'\text{-}0''$
34	$7'\text{-}5\frac{1}{4}''$	$7'\text{-}6\frac{11}{16}''$	$22'\text{-}8''$
35	$7'\text{-}7\frac{7}{8}''$	$7'\text{-}9\frac{5}{16}''$	$23'\text{-}4''$
36	$7'\text{-}10\frac{1}{2}''$	$8'\text{-}0''$	$24'\text{-}0''$
37	$8'\text{-}1\frac{1}{8}''$	$8'\text{-}2\frac{11}{16}''$	$24'\text{-}8''$
38	$8'\text{-}3\frac{3}{4}''$	$8'\text{-}5\frac{5}{16}''$	$25'\text{-}4''$
39	$8'\text{-}6\frac{3}{8}''$	$8'\text{-}8''$	$26'\text{-}0''$
40	$8'\text{-}9''$	$8'\text{-}10\frac{11}{16}''$	$26'\text{-}8''$
41	$8'\text{-}11\frac{5}{8}''$	$9'\text{-}1\frac{5}{16}''$	$27'\text{-}4''$
42	$9'\text{-}2\frac{1}{4}''$	$9'\text{-}4''$	$28'\text{-}0''$
43	$9'\text{-}4\frac{7}{8}''$	$9'\text{-}6\frac{11}{16}''$	$28'\text{-}8''$
44	$9'\text{-}7\frac{1}{2}''$	$9'\text{-}9\frac{5}{16}''$	$29'\text{-}4''$
45	$9'\text{-}10\frac{1}{8}''$	$10'\text{-}0''$	$30'\text{-}0''$

Note: When standard brick is used with concrete block, the mortar joints will be adjusted so that three courses of brick lay up to exactly 8 in., to match the concrete block lay-up.

FIGURE 4-8 (cont.)

Estimating Bricks Using the Volume Method

The procedure for using this method involves determining the number of bricks per cubic foot of wall. This may be done by dividing 1728, the number of cubic inches in 1 CF, by the volume, in cubic inches, of one brick plus its bed and end mortar joints. For a standard brick, the volume in cubic inches is:

$$\text{Volume} = 2.625 \times 4 \times 8.375 = 87.94 \text{ cubic inches}$$

Therefore, the number of standard bricks per cubic foot equals:

$$\text{Number of standard bricks} = \frac{1728}{87.94} = 19.65 \text{ bricks per CF}$$

It should be noted that the same result may be obtained by taking the number of bricks in 1 SF of one wythe and multiplying that value by three wythes, or 12 in. The result is, for a standard brick: 6.55×3, or 19.65 bricks.

CONCRETE BLOCKS

Concrete blocks are used for walls, partitions, pilasters, piers, and for backup material. When concrete blocks are used as a backup material for standard bricks, the mortar joints must be adjusted so that three courses of bricks lay up to equal exactly 8 in. This point is illustrated in Figure 4-9.

Concrete blocks are available in heavyweight and lightweight masonry units. The hollow heavyweight units weigh between 40 and 45 lb apiece. They are made with cement, sand, and various aggregates, including gravel, crushed rock, and slag. The

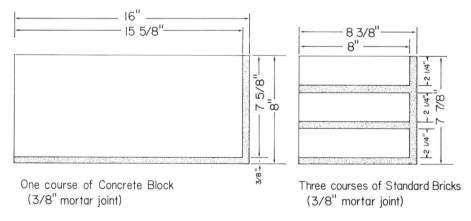

FIGURE 4-9 Mortar Adjustment Required to Match Standard Bricks and Concrete Blocks

lightweight units weigh between 25 and 35 lb apiece. They are made from cement, coal cinders, shale, clay, slag, and natural lightweight aggregates.

Although the actual size of a concrete stretcher block is $7\frac{5}{8}$ in. in width, $7\frac{5}{8}$ in. in height, and $15\frac{5}{8}$ in. in length, the nominal size is $8 \times 8 \times 16$. Other widths of concrete blocks are 2 in., 4 in., and 12 in., depending on construction needs. Mortar joints may be $\frac{1}{4}$ in., $\frac{1}{2}$ in., or other thicknesses; but the most common thickness is $\frac{3}{8}$ in. When a $\frac{3}{8}$-in. mortar joint is used, one course of concrete blocks lays up to approximately three courses of standard bricks (see Figure 4-9).

Figure 4-10 shows shapes and sizes for various masonry units, including corner, stretcher, double corner or pier, beam or lintel, joist, bull nose, and jamb concrete blocks.

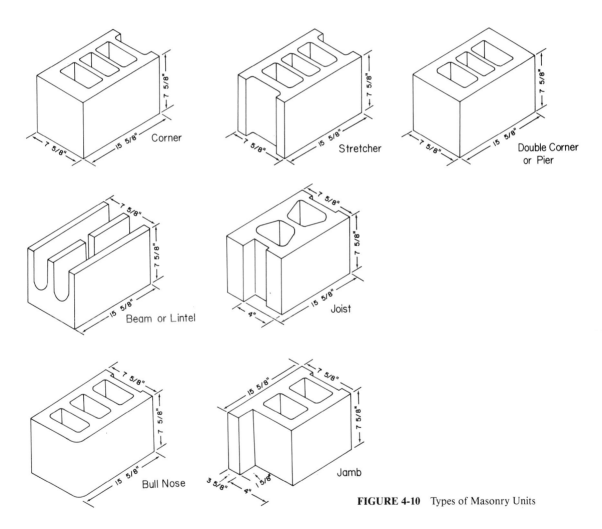

FIGURE 4-10 Types of Masonry Units

The take-off of concrete blocks is very similar to a take-off of bricks. The same procedures are used; only the size of the masonry unit is changed with concrete blocks.

Estimating Concrete Blocks Using the Wall Area Method

In using the wall area method, determine the number of concrete blocks in 1 SF of wall surface by dividing 144 by the product of 8×16, or:

$$\text{Number of concrete blocks per SF of wall} = \frac{144}{8 \times 16} = 1.125 \text{ blocks per SF}$$

Estimating Concrete Blocks Using the Course Method

To find the number of concrete blocks per course, divide the length of the wall, in inches, by the length of the block plus its mortar joint.

$$\text{Number of concrete blocks per course} = \frac{\text{length of wall (in.)}}{\text{length of block } + \text{ mortar joint (in.)}}$$

The number of courses is found by dividing the height of the wall, in inches, by the height of one block plus its mortar joint, or:

$$\text{Number of courses} = \frac{\text{height of wall (in.)}}{\text{height of block } + \text{ mortar joint (in.)}}$$

Using the same example of the brick wall which was 28 ft 7 in. long by 8 ft 9 in. high, the concrete blocks required as a backup for that wall are figured as follows:

$$\text{Number of concrete blocks per course} = \frac{28.58 \times 12}{16} = 21.43, \text{ or } 21\tfrac{1}{2} \text{ blocks per course}$$

$$\text{Number of courses} = \frac{8.75 \times 12}{8} = 13.12, \text{ or } 13 \text{ courses}$$

Total number of concrete blocks required equals 21.5×13, or 280 blocks. If a 5% waste factor is called for, the total number of blocks becomes 280×1.05, or 294 concrete blocks.

Allowance must be made when other types of concrete blocks are used in a wall area. To figure the number of corner blocks, jamb blocks, joist blocks, and so on, it is necessary to subtract the number of those blocks from the total number of stretcher blocks computed. For example, in figuring a requirement for 280 stretcher blocks in a wall 28 ft 7 in. long and 8 ft 9 in. high, corner blocks are needed at each end of the wall. Since the wall is 13 courses high, 26 corner blocks are needed. This amount is subtracted from 280, giving a result of 254 stretcher blocks, plus a waste factor, or 254 times 1.05, giving 269 stretcher blocks. A waste factor should also be applied to the 26 corner blocks, giving a result of 27 corner blocks. The same procedure may be used to determine the numbers of other types of concrete blocks.

REINFORCEMENT

Reinforcement is required in most load-bearing walls and partitions to prevent stress cracking. Two longitudinal wires which are joined together by transverse wires are placed in every second or third course. In figuring this reinforcement, a constant can be developed based on the amount of wire required per square foot of wall. For example, if wire reinforcement is placed in every second course, it would take 1.33 LF, or 16 in., the length of one block, of reinforcement for every 1.77 SF, or 1.33 × 1.33, the area covered. Therefore, to find the number of linear feet of reinforcement required per square foot, divide 1.33 by 1.77, giving 0.75 LF. To provide for lapping of the reinforcement at wall corners, add a waste factor of 10%, giving 0.83 LF of reinforcement per square foot of wall.

Using the same procedure, constants may be developed for reinforcement spaced every third course or for other spacings. Figure 4-11 gives the results for spacings of reinforcement every second and every third course.

Number of Courses between Reinforcement	Joint Reinforcement per Block LF	Joint Reinforcement per SF
2	0.74	0.83
3	0.48	0.55

Note: Figures include a 10% lap factor.

FIGURE 4-11   Joint Reinforcement Constants

In estimating the number of masonry ties required in a wall, there are several methods which may be used. One method is that described in Chapter 3 for figuring concrete form ties. Another method involves developing constants for various masonry tie spacings. These constants are simply the approximate number of ties per square foot of wall based on a particular spacing of ties. Figure 4-12 gives constants for four possible spacings.

SPACING OF MASONRY TIES		
Vertically (in.)	Horizontally (in.)	Tie Constant (number of ties required per SF of wall area)
12	12	1.00
12	24	0.50
16	32	0.28
24	24	0.25

Note: No waste factor included.

FIGURE 4-12   Masonry Tie Constants

The purpose of masonry ties is to give lateral strength to a masonry wall by tying it to another masonry wall or a wall of another material. A good example of a wall of another material is wood frame construction. In this type of construction, wall ties are used to fasten the stud wall framing to the bricks or other masonry units, giving them additional lateral strength.

In some specifications, the spacings of wall ties are given as 16 in. on centers vertically, and 32 in. on centers horizontally. For this type of specification, the constant is figured as follows:

$$\text{Number of square feet for each tie} = 1.33 \text{ ft} \times 2.67 \text{ ft} = 3.56 \text{ SF}$$

$$\text{Number of ties per square foot} = \frac{1}{3.56} = 0.28$$

Other constants may be developed in the same way for other spacings. The two methods for computing ties mentioned above give approximately the same results for the same spacings. For example, the wall form measuring 50 ft long and 8 ft high, mentioned in Chapter 3, required 113 ties using that method. Using the same dimensions and the same spacings, the constant in Figure 4-12 gives the following results:

$$50 \times 8 \times 0.28 = 112 \text{ ties}$$

MORTAR

The mortar required in masonry construction is based on the number of bricks, blocks, or stones to be laid. A certain volume of mortar, given in cubic feet or number of sacks, will be needed to lay 1000 bricks of a given size and type. This amount may be computed for any type of bricks. For example, a standard brick using a $\frac{3}{8}$-in. end joint and a $\frac{3}{8}$-in. bed joint will require approximately 8.8 CF of mortar per 1000 bricks. A standard concrete block, using $\frac{3}{8}$-in. mortar joints, will require approximately 19 CF of mortar per 1000 blocks. Since one bag of mortar equals 1 CF, the number of cubic feet required will also be the number of bags required. A waste factor is usually applied when job conditions warrant.

Type of Bricks	$\frac{3}{8}$-in. Joint	$\frac{1}{2}$-in. Joint
Standard	8.8 CF	11.9 CF
Modular	8.1	10.3
Roman	10.6	13.7
Norman	11.2	14.3

Note: Depending on job conditions, craftsmen's experience, weather conditions, and so on, a waste factor of 5 to 15% should be applied to the figures above.

FIGURE 4-13   Mortar Required in Cubic Feet per 1000 Bricks

Although the mortar required for the setting of stone will vary with the size of the stone, the width and thickness of the bed, and so forth; a good approximation is 4 to 5 CF of mortar per 100 CF of stone. Figure 4-13 gives mortar quantities, in cubic feet, per 1000 masonry units of standard, modular, Roman, and Norman bricks for $\frac{3}{8}$- and $\frac{1}{2}$-in. mortar joints.

STONE

When stone is used as a veneer or facing for a wall, the working drawings or specifications usually call for an average thickness for that stone, and the cost is based on so much per square foot for that thickness. However, when stone is used as rubble masonry for solid walls, the take-off is usually figured as so many cubic yards. Stone may be estimated by the linear foot when it is used as trim around openings, sills, band courses, and so on.

Other items that must be considered in a stone take-off are anchors, dowels, clamps, and other accessories involved in setting stone. Stone will vary widely as to cost and type depending on the geographical location and availability. Some of the principal types of stone are rubble stone, ashlar stone, and cut stone.

Rubble stone is irregularly shaped and comes in thicknesses ranging from 2 to 8 in. Ashlar stone has a flat surface which is usually rectangular with smooth beds and joints. Cut stone is cut and delivered to the site ready to set in place. Granite, limestone, sandstone, marble, and slate are some of the types of cut stone available in the United States.

WORK EXERCISE 10: CONCRETE AND MASONRY (CSI DIVISIONS 3 and 4)

Concrete and Reinforcement Bricks and Mortar

Objective: Using the elevation, plan, and cross sections of the retaining wall shown in Figure 4-14, make take-offs for the following items.

- *Concrete:* Determine the cubic yards of concrete required for the footing and retaining wall using a waste factor of 3%. Note that in the 40-ft section of the wall, the height varies from 8 to 15 ft. Also note that the thickness of the 40-ft section varies from 8 in. at the top to 1 ft 4 in. at the bottom. The thickness of the 96-ft section of the wall varies from 8 in. at the top to 1 ft 11 in. at the bottom of the wall.
- *Reinforcement:* Section A-A shows No. 7 rebars running horizontally in the 40-ft section of the wall, and No. 3 rebars running horizontally in the footing. The length of the vertical steel is given in the cross sections. Use a waste factor of 8% for reinforcement.
- *Masonry:* Using modular bricks which measure 4 in. $\times 2\frac{2}{3}$ in. $\times 8$ in., determine the number of bricks required for the face of the retaining wall which is 136 ft 0 in. in length.

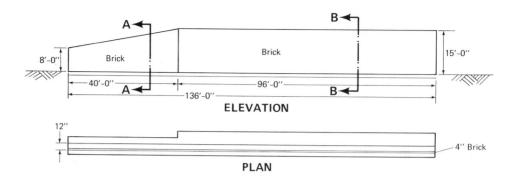

ELEVATION

PLAN

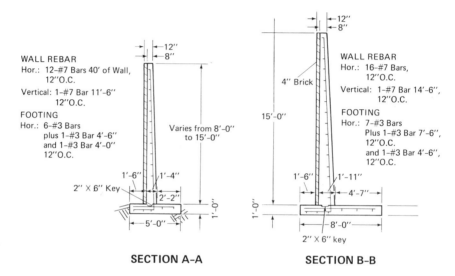

SECTION A-A   SECTION B-B

FIGURE 4-14   Retaining Wall

Use a waste factor of 2%, and assume that 100 SF of wall area requires 675 modular bricks.

• *Mortar:* Estimate the number of sacks or bags (1 CF equals one bag) of mortar required for this job if every 1000 bricks require 8.1 CF of mortar. For this take-off, use a waste factor of 10%.

Compute the following:

1. Cubic yards of concrete for footings
2. Cubic yards of concrete for retaining wall
3. Pounds of No. 7 rebar reinforcement
4. Pounds of No. 3 rebar reinforcement

5. Number of modular bricks

6. Number of bags of mortar (1 bag = 1 CF)

WORK EXERCISE 11: CONCRETE AND MASONRY (CSI DIVISIONS 3 and 4)

Concrete and Reinforcement, Bricks and Mortar

Objective: Using Figures 4-15 and 4-16, make a take-off of the concrete, reinforcement, cushion sand, modular bricks, and mortar required for a small building.

Procedure

In studying the figures, note that the modular brick wall is two wythes wide and 9 ft 0 in. high. Assume a mortar joint of $\frac{3}{8}$ in. thickness.

      Determine the stretch-out length, and calculate the volume of concrete needed for the footings, basement wall, and the number of pounds of reinforcement required. In figuring the net area of the walls, subtract rough openings for windows which are 2 ft wide and 3 ft high in the cellar, and 2 ft 6 in. × 6 ft 6 in. on the first floor. Rough openings for the two exterior doors are 3 ft × 7 ft.

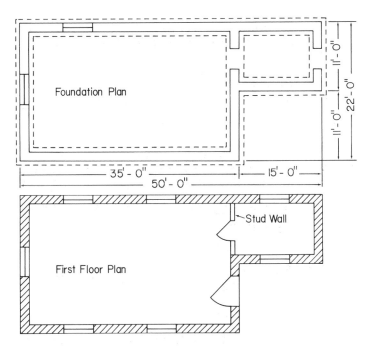

FIGURE 4-15 Foundation and First-Floor Plans

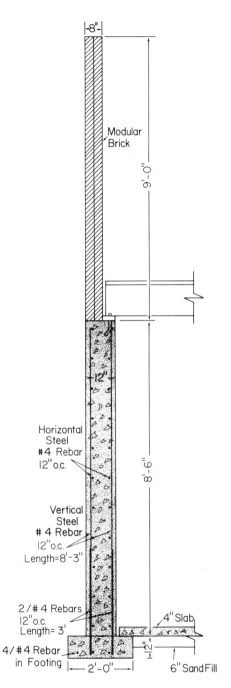

FIGURE 4-16   Wall and Footing Cross Section

Next, determine the cubic yards of concrete required for the 4-in. slab located in the basement, and the volume of the cushion sand of 6 in. thickness to be placed under the slab. Finally, determine the number of modular bricks and the number of bags of mortar required for a wall two wythes wide. (A wythe is defined as a vertical section of a wall, one masonry unit in thickness.)

Use waste factors as follows:

Concrete, 8%
Reinforcement, 8%
Cushion sand, 10%
Modular bricks, 2%
Mortar, 10%

Compute the following:

1. Cubic yards of concrete for footings
2. Cubic yards of concrete for basement wall
3. Pounds of No. 4 rebar in the footings and basement wall
4. Cubic yards of concrete for 4-in. slab
5. Cubic yards of cushion sand
6. Number of modular bricks
7. Number of bags of mortar (1 bag = 1 CF; figure 8.1 CF of mortar for every 1000 bricks laid)

WORK EXERCISE 12: MASONRY (CSI DIVISION 4)

Modular and Standard Bricks, Mortar, and Wall Ties

Objective: Make a take-off of masonry items for a residence consisting of two alternates, one using modular bricks, and another one using standard bricks. Refer to Figure 3-17 for the foundation dimensions, and to Figure 4-17 for a typical wall section for a small residence. The take-offs for the two alternates will include bricks, mortar, and wall ties.

Notes

1. For this house, the door openings have been figured as a total of 206 SF.
2. Window openings have been figured as a total of 236 SF.
3. Wall ties are spaced 16 in. on centers vertically and 32 in. on centers horizontally.
4. Modular bricks measure 4 in. $\times$ $2\frac{2}{3}$ in. $\times$ 8 in. with dimensions including mortar thickness.
5. Standard bricks measure $3\frac{3}{4}$ in. $\times$ $2\frac{1}{4}$ in. $\times$ 8 in. with dimensions not including the mortar thickness.
6. Refer to Figure 4-8 to determine the number of brick courses per wall height.
7. Refer to Figure 4-12 to determine wall tie constant.

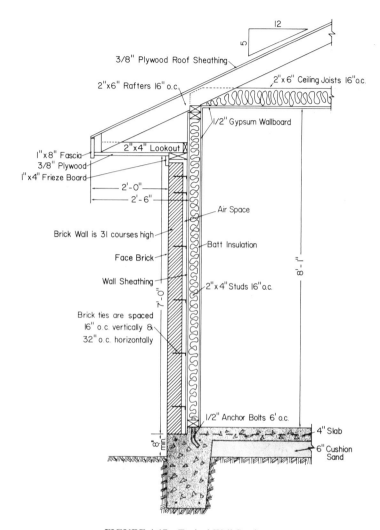

FIGURE 4-17 Typical Wall Section

8. Use waste factors as follows:

Bricks, 2%
Mortar, 10%
Wall ties, 5%

9. Refer to Figure 4-13 to determine the amount of mortar required.

Compute the following:

Alternate 1:

1. Number of modular bricks

2. Bags of mortar

3. Number of wall ties

Alternate 2:

4. Number of standard bricks

5. Bags of mortar

6. Number of wall ties

WORK EXERCISE 13: MASONRY (CSI DIVISION 4)

Concrete Blocks by Types and Mortar

Objective: Using Figures 4-18 and 4-19, determine the numbers of various types of concrete blocks in a small masonry building.

Procedure

Using the information below, determine the total number of concrete blocks that would be required if all the blocks in the four walls were stretcher blocks. Next, determine the numbers of special types of blocks, which include corner, bond beam, lintel, and door and window jamb blocks.

The total number of stretcher blocks actually required equals the total number of blocks in the four walls minus the total of all the special blocks.

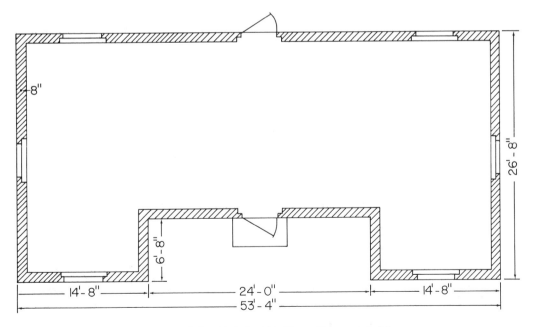

FIGURE 4-18 Plan View of Masonry Building

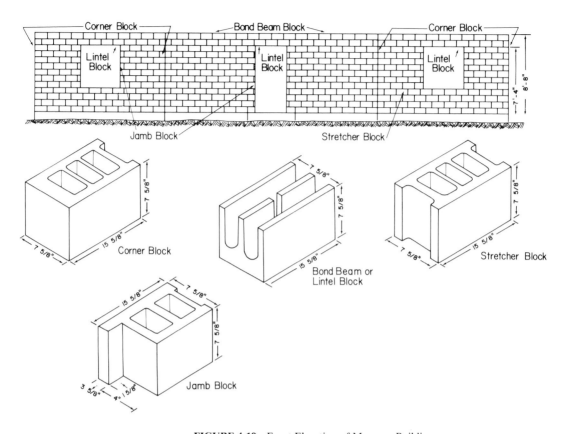

FIGURE 4-19   Front Elevation of Masonry Building

Notes

1. Use a waste factor of 5% for all concrete blocks.
2. Rough openings for windows are 4 ft 0 in. × 4 ft 8 in.
3. Rough openings for doors are 3 ft 4 in. × 7 ft 4 in.
4. In estimating the mortar, figure 19 CF of mortar for every 1000 concrete blocks laid.

Compute the following:

1. Number of corner blocks
2. Number of bond beam blocks
3. Number of lintel blocks
4. Number of full door and window jamb blocks
5. Number of half door and window jamb blocks
6. Number of stretcher blocks
7. Number of bags of mortar

WORK EXERCISE 14: MASONRY (CSI DIVISION 4)

Modular Bricks and Concrete Blocks

Objective: Using Figure 4-20, which shows the plan view and wall section of a small masonry building, estimate the quantity of modular bricks, concrete blocks, and mortar required.

Notes

1. Window openings are 4 ft 2 in. × 6 ft 0 in.
2. Door openings are 3 ft 6 in. × 7 ft 0 in.
3. Figure 8.1 CF of mortar for every 1000 bricks laid.
4. Figure 19 CF of mortar for every 1000 blocks laid.
5. Apply waste factors as follows:

$$\text{Bricks, } 3\%$$
$$\text{Concrete blocks, } 5\%$$
$$\text{Mortar, } 10\%$$

Compute the following:

1. Number of modular bricks
2. Number of concrete blocks
3. Number of bags of mortar

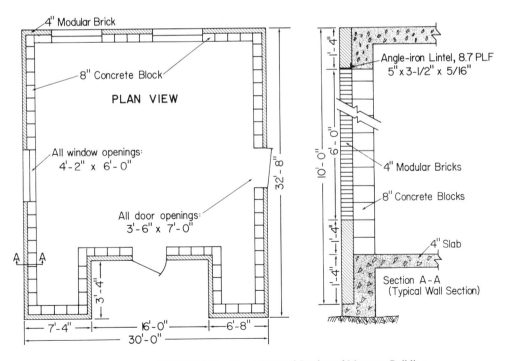

FIGURE 4-20   Plan View and Section of Masonry Building

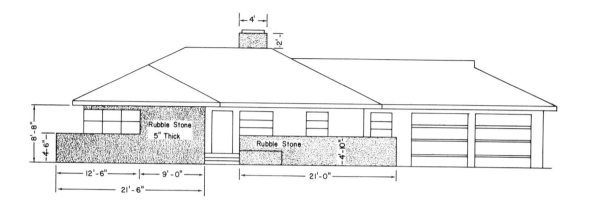

Front Elevation

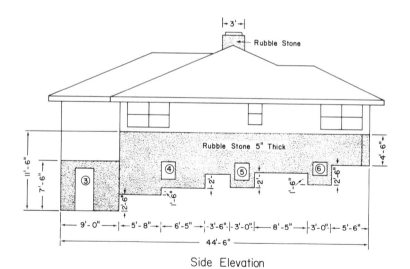

Side Elevation

FIGURE 4-21   Residence: Front and Side Elevations

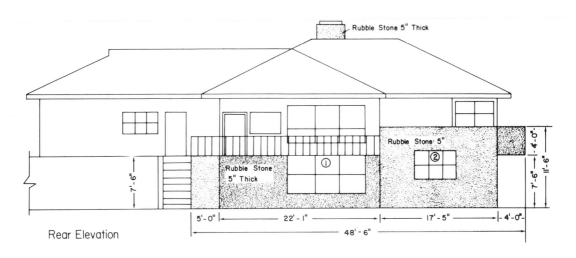

Rear Elevation

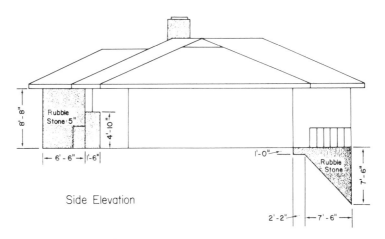

Side Elevation

FIGURE 4-22   Residence: Rear and Side Elevations

WORK EXERCISE 15: MASONRY (CSI DIVISION 4)

Stone and Mortar

Objective: Using Figures 4-21 and 4-22, which show four elevations of a residence, estimate the volume of rubble stone and mortar required.

Procedure

In making a stone take-off, it is first necessary to locate where the stone is to be installed. On the four elevations for this residence this has been done by shading the areas where the rubble stone is to be applied.

The next step is to calculate the net square footage of these areas, and then multiply that value by the thickness, in feet, of the rubble stone. In this exercise, the rubble stone is 5 in. thick, which converts to 0.42 ft thick.

In determining the net square footage of the stone area, window and door openings must be subtracted from the gross area. The total volume of the stone, in cubic feet, is obtained by multiplying the net area of the stone by its thickness.

Notes

1. The thickness of the rubble stone is 5 in.

2. Assume that 4.5 CF of mortar will be required for every 100 CF of rubble stone set.

3. Use a waste factor of 5% for the stone and 10% for the mortar.

4. Rough openings for windows and doors are as follows:

Item	Rough Openings
Window 1	11′-0″ × 5′-6″
Window 2	6′-6″ × 4′-6″
Door 3	2′-8″ × 7′-6″
Window 4	2′-6″ × 2′-8″
Window 5	2′-6″ × 2′-8″
Window 6	2′-0″ × 2′-0″

Compute the following:

1. Cubic feet of rubble stone

2. Bags of mortar

5

Metals

Upon completion of this chapter, the student will be able to:

- Make a partial take-off of structural steel members in an office building.
- Make a take-off of metal decking.

GENERAL

In preparing a take-off of metals involved in a construction project, the estimator must examine the plans and working drawings closely in order to determine the quantities of the various items. These items include structural steel, metal open web joists, junior steel beams, and miscellaneous shapes. The estimator must determine which items are to be subcontracted and what is included in that work. For example, structural steel is usually a subcontract item, but many structural steel firms make a practice of excluding standard open web joists and long-span joists. On the other hand, there are steel fabricators who specialize in open web joists and long-span joists. It is the estimator's responsibility to find out which situation applies and to prepare the estimate accordingly.

Since the general contractor's estimator often prepares a structural steel take-off only for the purpose of checking and evaluating the subcontractors' bids, the amount of detail for those take-offs may be greatly reduced. In many situations, take-offs for structural steel, steel bar joists, metal decking, and miscellaneous items form the major part of the estimate. Costs for structural steel and steel bar joists are

79

FIGURE 5-1 Erecting Structural Steel

usually determined on a weight basis. Metal decking costs are based on so much per square foot.

One of the principal costs of structural steel on a project is the cost of erection. Figure 5-1 shows structural steel members being erected on a multistory office building. In developing equipment costs, the estimator must refer to published equipment rates and rely on past experience.

STRUCTURAL STEEL

Although there are a number of items to be considered in estimating the cost of structural steel on a project, most of these items are directly related to the quantity required. Therefore, the first step is to make a quantity take-off of the various steel shapes in terms of pounds or tons. Figure 5-2 shows the manufactured beam shapes, angles, tees, plates, built-up sections, and tubing which are available. The steel shapes are designated by a group symbol which is followed by a nominal depth and weight per foot of length. For example, the designation $W10 \times 49$ indicates a wide-flange shape, approximately 10 in. in depth and weighing 49 lb per linear foot.

The structural steel shape or group symbols, their names, and principal characteristics are as follows:

W   Wide-flange shapes: parallel inner and outer flange surfaces

M   Miscellaneous shapes: special lightweight shapes having a profile similar to W shapes

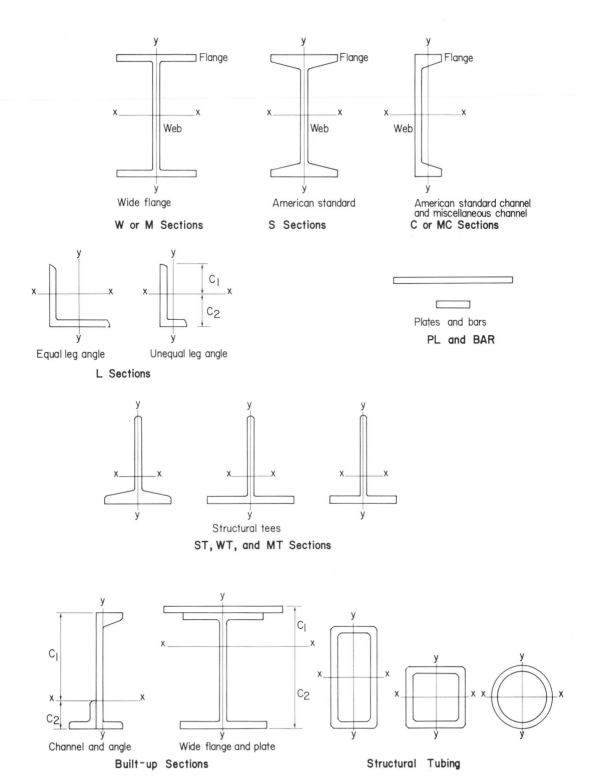

FIGURE 5-2 Manufactured Beam Shapes

81

S	American standard beams: webs thicker and flanges narrower than the W shapes
C	American standard channels: C-shaped with tapered inner surfaces on flanges
MC	Miscellaneous channels: special-purpose shapes having the same profile as C shapes
L	Angles: angle legs, either equal or unequal in length, set at right angles to each other
ST, WT, and MT	Structural tees: S, W, and M shapes split at middepth
PL & BAR	Plates and bars
TS	Structural tubing: square, rectangular, or circular in shape

These symbols supersede older designations which may be encountered on some blueprints and drawings. For example, WF or wide-flange beams are now called W sections. Formerly referred to as I beams, the new designation for this item is S, or standard beam. Junior beams, such as 6 Jr. 4.4, are lightweight beams used in light-occupancy buildings such as stores, apartments, residences, and schools. The new designation for this type of beam is an M shape, M6 × 4.4, which means that a steel beam is called for with a 6-in. depth and a weight of 4.4 lb per linear foot.

METAL JOISTS

Standard open web and long-span steel joists are made by welding top and bottom chord members to round bars or angle web members as shown in Figure 5-3.

FIGURE 5-3 Long-Span Steel Trusses

Steel joist construction is used to great advantage in all types of light-occupancy buildings. To use steel joists in "fireproof" construction, it is necessary to top them with a minimum 2-in.-thick reinforced concrete slab, and a fire-resistant ceiling of gypsum board beneath the joists.

Steel joists are generally sold directly to the contractor by the manufacturer. Therefore, when the general contractor is preparing the steel joist take-off for an estimate, copies of the job plans may be sent to the joist manufacturer for evaluation. The manufacturer may then furnish additional information concerning joist accessories such as clips, bridging, and ceiling extension rods.

METAL DECKING

Steel roof decking comes in varying gauges of strip steel and in various widths and lengths. Most decking sections are $1\frac{1}{2}$ in. deep, but they are also available in 2- and $2\frac{1}{2}$-in. depths. For spans from 7 ft 0 in. to 8 ft 6 in., 18-gauge decking is used, and for spans 6 ft 0 in. to 7 ft 6 in., 20-gauge is usually used. When metal decking is placed on top of steel joists, the metal decking is usually welded to the joists.

Spans up to 32 ft 0 in. may be achieved through the use of "long-span" decking. Long-span decking is made in $4\frac{1}{2}$-, 6-, or $7\frac{1}{2}$-in.-deep pans which are 12 in. wide and made of 14-, 16-, 18-, or 20-gauge steel. One size of metal decking is shown in Figure 5-4.

FIGURE 5-4   Metal Decking (Courtesy of Wheeling Corrugating Company, a Division of Wheeling-Pittsburgh Steel Corporation, Wheeling, W. Va.)

FIGURE 5-5   Steel Framing (Courtesy of Wheeling Corrugating Company, a Division of Wheeling-Pittsburgh Steel Corporation, Wheeling, W. Va.)

MISCELLANEOUS STEEL

Under the miscellaneous category are such items as light-gauge framing, stairs, railings, porch columns, and others.

Light-gauge framing systems consist of studs, joists, and accessories for the steel framing of buildings up to four stories in height. The advantages of using these metal members include less dead load, uniform fabrication, reduced on-the-job storage space, strength, and low cost. Figure 5-5 shows the framing of a wall section using metal studs. These studs are available in $2\frac{1}{2}$-, $3\frac{5}{8}$-, 4-, and 6-in. depths. They come in various gauges, including 12, 14, 16, 18, and 20 gauge.

Light-gauge joists are available in 6-, 8-, 10-, and 12-in. depths, and in 12-, 14-, 16-, 18-, and 20-gauge material. These joists and studs are estimated as so much cost per linear foot, depending on the type and weight of each item.

In estimating metal stairs (Figure 5-6) the cost is usually based on the number of risers, with stringers, treads, nosing, and railing reflected in that cost.

Pipe railings may be estimated as so much per linear foot depending on the material, number of horizontal runs, and height and spacing of uprights.

Porch columns are estimated individually based on the design, type, and size.

FIGURE 5-6 Metal Stairs

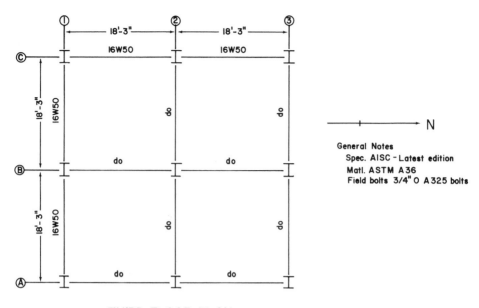

THIRD FLOOR PLAN

Finished Floor Elevation 36'-4"
Top of steel 4" below Finished Floor

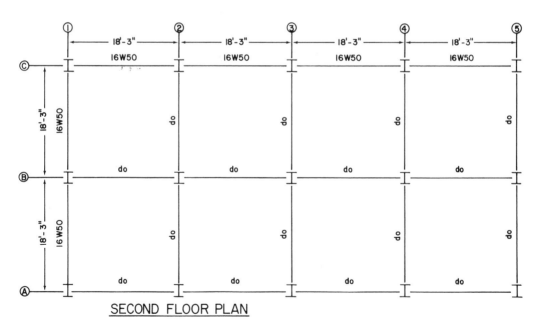

SECOND FLOOR PLAN

Finished Floor Elevation 24'-4"
Top of steel 4" below Finished Floor

FIGURE 5-7   Floor Framing Plans

WORK EXERCISE 16: METALS (CSI DIVISION 5)

Structural Steel

Objective: Using the floor framing plans in Figure 5-7, make a partial structural steel take-off of the second and third floors of a small office building.

Notes

1. Estimate the approximate weight of the wide-flange beams which frame into the columns (no waste factor).
2. Estimate the approximate square footage of metal decking that will be required for these two floors (use a 5% lap factor).

Compute the following:

1. Number of tons of 16W50 beams required
2. Number of square feet of metal floor decking required

WORK EXERCISE 17: METALS (CSI DIVISION 5)

Door and Window Lintels

Objective: Using Figure 5-8, determine the number of pounds of angles that are required for door and window lintels in a small masonry building.

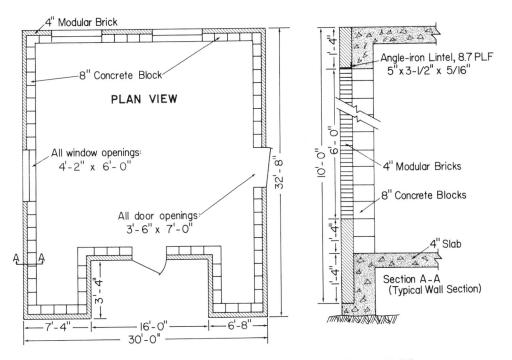

FIGURE 5-8 Plan View and Section of Masonry Building

Notes

1. Each lintel is to bear a minimum of 6 in. on each side of the openings.
2. Use a waste factor of 2% for the lintels.
3. Rough openings for windows are 4 ft 2 in. × 6 ft 0 in.
4. The rough opening for the door is 3 ft 6 in. × 7 ft 0 in.
5. All lintels are 5 in. × $3\frac{1}{2}$ in. × $\frac{5}{16}$ in. and weigh 8.7 PLF.

1. Compute the total pounds of lintels required.

6

Wood and Plastics

OBJECTIVES

Upon completion of this chapter, the student will be able to:

- Locate and identify the various construction items involved in estimating the components of a wood frame structure.
- Make a rough carpentry take-off of quantities required for underpinning, floor systems, walls and partitions, ceilings, and roof system for a house.
- Make a finish carpentry take-off of quantities required for siding, frieze boards, fascia, soffits, cabinets and millwork, moldings, paneling, and stairs for a house.

GENERAL

Carpentry take-offs may be divided into rough carpentry and finish carpentry under Division 6 of the Construction Specifications Institute MASTERFORMAT. Included in the rough carpentry category are those construction items that make up the framing and sheathing of a structure. In comparing a building structure to the human body, the wood framing is similar to the bones of the skeleton, and the sheathing is similar to the skin. The framework gives a structure stability and strength; and it includes sills, posts, girders, plates, studs, floor and ceiling joists, and rafters. The sheathing ties the whole structure together as a unit, and it forms a base for the thermal and moisture protection materials. Sheathing consists of plywood or other similar materials.

Finish carpentry is important from an appearance and utility standpoint; and it includes siding, cornice work, cabinets and millwork, moldings, trim, paneling, and other specialties.

ROUGH CARPENTRY

Dimension lumber is framing lumber that is 2 in. or more in thickness. It is normally used for the structural members of a wood frame building. Although other materials are sometimes used, plywood is the most commonly used material for the sheathing of a wood frame building. As a convenience in making a rough carpentry take-off, a building may be divided into five parts: (1) underpinning, (2) floor system, (3) walls and partitions, (4) ceilings, and (5) roof system.

Underpinning

In order for the floor joists and floor system of a building to be supported, an underpinning is required. This underpinning consists of posts, girders, and mudsills, as shown in Figure 6-1. In this type of underpinning, posts sit on top of pier blocks and support girders. To estimate posts, multiply their average lengths by the number of posts, and convert that value to board feet.

The size and length of girders will be shown on the working drawings with their spans depending on loads and grades of lumber. An estimate of girders may be made by taking off their total lengths, multiplying by the number of pieces required to form one girder, and converting that value to board feet.

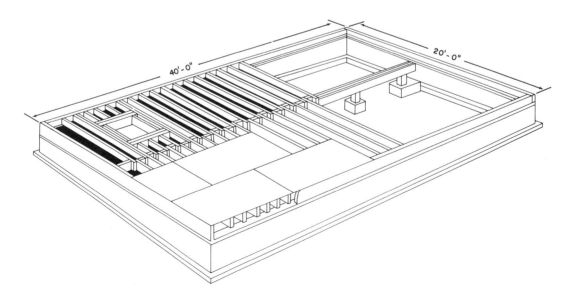

FIGURE 6-1 Typical Wood Floor System

Mudsills are plates that are in contact with concrete or masonry foundation walls. Mudsills may be of random lengths, and are estimated by determining their total lengths, and converting that value to board feet. An easy way to figure the length of a mudsill is by using the stretch-out-length of the perimeter beam to which the mudsill is fastened.

Floor System

Included in this category are floor joists, header joists, subflooring, and underlayment. Floor joists (Figure 6-2) are horizontal framing members which are supported at one end by a girder and at the other end by a mudsill. On second-floor systems, floor joists are supported at one end by an exterior wall top plate, and at the other end, by an interior partition top plate.

A take-off of floor joists in a floor system may be made by first determining the number of joists required of a specific size and length. The next step is to convert that quantity of floor joists into board feet.

Lumber sizes have been standardized for convenience in ordering and handling. They come in even-numbered lengths, such as 6, 8, 10, 12, 14, 16, 18, and 20 ft. The nominal widths are 2, 4, 6, 8, 10, and 12 in. The nominal thicknesses are 1, 2, and 4 in. The actual width and thickness of dimension lumber is less than the nominal width and thickness. For example, a 2 × 4 is the nominal dimension for a piece of lumber which actually measures $1\frac{1}{2} \times 3\frac{1}{2}$.

A board foot is a common unit of measure for lumber. It represents a piece of lumber having a flat surface area of 1 SF, and a nominal thickness of 1 in. To deter-

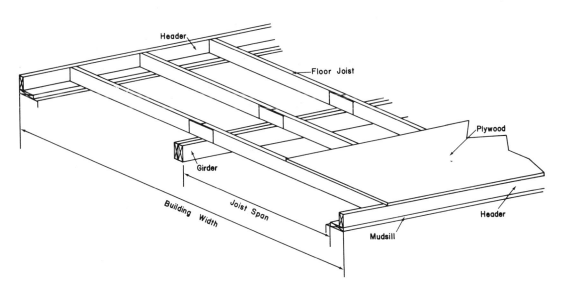

FIGURE 6-2   Parts of a Wood Floor System

mine the number of board feet in one or more pieces of dimension lumber, the following procedure may be followed:

$$\text{Board feet (BF)} = \frac{\overset{\text{number}}{\underset{\text{of pieces}}{}} \times \overset{\text{nominal}}{\underset{\text{thickness (in.)}}{}} \times \overset{\text{nominal}}{\underset{\text{width (in.)}}{}} \times \text{length (ft)}}{12}$$

Application of this formula may be shown by the use of an example. To find the
number of board feet in six pieces of 2 × 8's which are 20 ft long, simply plug in those
numbers, and divide by 12.

$$\text{BF} = \frac{6 \times 2 \times 8 \times 20}{12} = 160 \text{ BF}$$

The answer for this example, 160 BF, may also be written as 0.160 MFBM. MFBM
stands for 1000 ft, board measure, and it is a unit often used in the construction
industry.

The number of floor joists is determined by dividing the length of the floor
system, measured perpendicular to the direction of the joists, by the spacing of the
joists, and then adding one additional joist to that quotient. For example, for joists
spaced 16 in. on centers, the length of the floor system would be divided by 1.33 ft
(16 in.), and then 1 would be added to that value. If the floor joists are spaced 24 in.
on centers, the floor length would be divided by 2, and then 1 would be added. The
addition of one joist is for a starter joist. It will also be necessary to add one extra
floor joist for every partition that runs parallel to the direction of the floor joists.

Instead of dividing by 1.33 and 2 in the example above, the same results can be
obtained by multiplying the floor length by $\frac{3}{4}$ for 16-in. spacing, and by $\frac{1}{2}$ for 24-in.
spacing. Factors to multiply times floor lengths to obtain the number of joists for
other spacings are listed below.

Joist Spacing (in.)	Factor
12	1
16	$\frac{3}{4}$
24	$\frac{1}{2}$
32	$\frac{3}{8}$

To figure the number of joists required for the floor system in Figure 6-1, for
any of the joist spacings above, multiply the length, 40 ft, by the factor for that particular spacing, and then add 1. For example, for floor joists spaced 16 in. on centers,
the calculation is as follows:

$$\text{Number of joists in one bay} = (40 \times \tfrac{3}{4}) + 1 = 31 \text{ joists}$$

$$\text{Number of joists in two bays} = 2 \times 31 = 62 \text{ joists}$$

This answer, 62 joists, is based on single joists throughout the floor system without openings. Extra joists must be added where partitions run parallel to the joists and where the joists are doubled. Depending on the size of openings, the total number of joists will be reduced by an appropriate number.

Header joists are dimension lumber members of the same size as the floor joists. They are usually set edgewise on top of the mudsill, as shown in Figure 6-2. The linear feet of header joists required is determined, and then that value is converted to board feet. Since header joists may be of random length, any convenient even-length foot dimension may be specified.

Various building materials are used as sheathing, or decking, over floor joists; but the most commonly used sheathing material is plywood. Depending on design loads and spacing of floor joists, plywood sub-flooring may vary in thickness from $\frac{3}{8}$ to $1\frac{1}{8}$ in. For greatest strength and stiffness, plywood panels should be applied across floor joists, with the face grain direction of the plywood perpendicular to the direction of the floor joists. To takeoff the number of standard-size plywood panels required for a floor system, determine the total area of the floor, in square feet, and then divide by 32. The width and length of a standard size plywood panel is 4 ft × 8 ft, or 32 SF of surface area.

Sometimes a combination subfloor and underlayment grade of plywood is used for a floor system. When this is done, the take-off procedure is exactly the same. The only difference is in the cost of the material because the grade of the plywood will be higher compared to just subfloor grade.

This point may be illustrated by an example. To make a take-off of the number of plywood panels required for the floor system in Figure 6-1, the procedure is to take the area and divide it by 32. This is true regardless of whether the plywood is CDX sheathing, a subfloor grade, or whether the plywood is a higher subfloor/underlayment grade. In either case:

$$\text{Number of panels} = \frac{40 \times 20}{32} = 25 \text{ panels}$$

Walls and Partitions

Walls may be defined as exterior components receiving exterior finishing on one side. Partitions may be defined as interior components receiving interior finishing on both sides. Both exterior walls and interior partitions are made up of top and bottom plates, studs, bracing, headers, and firestops. In making a wall and partition take-off, determine the plate-line length for both the exterior walls and the interior partitions. The plate-line length of the exterior walls may be computed by using the stretch-out-length method. For use later, the lengths of wall and partition openings should be determined.

In making a take-off of walls and partitions, various aspects of the types of construction must be noted. For example, studs used in Western or platform construction are shorter than those used in balloon-type construction. Figure 6-3 shows how the studs in the Western type run from the sole plate to the top plate. In Figure 6-4, the studs of the balloon-type run the full height from the sill to the top plate of the second story.

Plates are horizontal framing members which form the bottom, or sole, and the top of a wall or partition. Normally, they are 2×4's of random lengths, with one plate on the bottom and two plates on top. When special widths are required for plumbing partitions, 2×6 or 2×8 plates must be figured separately from the 2×4 plates.

When one bottom plate and two top plates are used in a wall and partition combination, the total plate-line length is figured by multiplying the total plate-line length by 3. Normally, the widths of the openings are not taken out if those openings are less than 3 ft.

Studs are vertical framing members spaced to form the framework of a wall or partition. They are usually 2×4's unless, as with plates, they are used in a plumbing

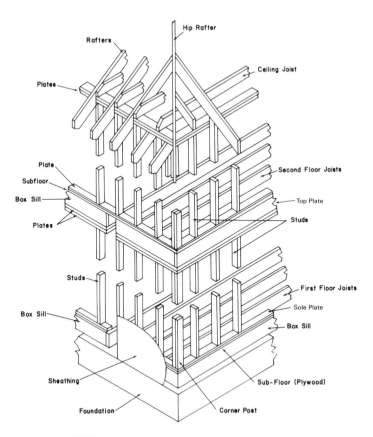

FIGURE 6-3   Western or Platform-Type Framing

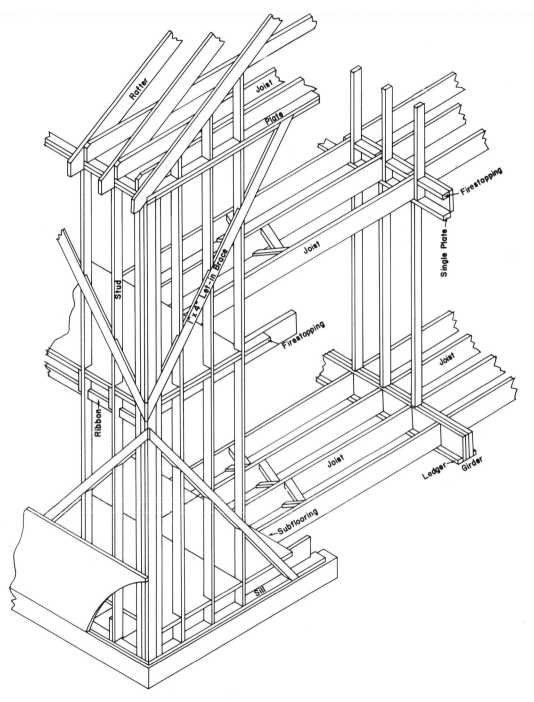

FIGURE 6-4   Balloon-Type Framing

wall which requires a greater wall thickness. When studs are spaced 16 in. on centers, there are two principal methods for making a stud take-off. The first method is the more accurate of the two, but the second method is faster.

Method 1 for stud take-offs. Since studs are doubled at openings and tripled at corners and tees (the intersection of a partition and a wall), studs may be figured as follows:

1. Divide the total length of the walls and partitions by the stud spacing, or multiply the plate-line length by $\frac{3}{4}$ for studs spaced 16 in. on centers.
2. Figure one extra stud for each corner shown on the floor plan.
3. Figure one extra stud for each space between the tees, plus one stud for each space between the tees and corners.
4. Figure two extra studs for each window or door opening of 3 ft or less as shown on the floor plan.
5. The total number of studs required is obtained by adding the values of these four items.

Method 2 for stud take-offs. This method is an approximation which may be used when the exterior wall openings are between 15 and 20% of the heated area, and when the studs are spaced 16 in. on centers. Under these conditions, the stud take-off procedure is as follows:

Determine the total length of the walls and partitions, and multiply that value by 1. In other words, the total plate-line length equals the number of studs required. Although the studs are actually spaced 16 in. on centers, this method figures one stud per foot of wall or partition length. The additional studs thus gained are available for corners, tees, and openings. It should be noted that although this method is faster and simpler than Method 1, it is not as accurate.

A comparison of the two methods will illustrate this point. Using the floor plan in Figure 6-5, the stud take-offs, using the two methods, is as follows:

Method 1 for stud take-offs

1. Total LF of walls and partitions	$225 \times \frac{3}{4}$ =	168.8
2. Number of corners	10×1 =	10
3. Number of spaces	22×1 =	22
4. Number of openings	23×2 =	46
5. Total		246.8, or 247 studs

Method 2 for stud take-offs

1. Total LF of walls and partitions 225×1 = 225   studs

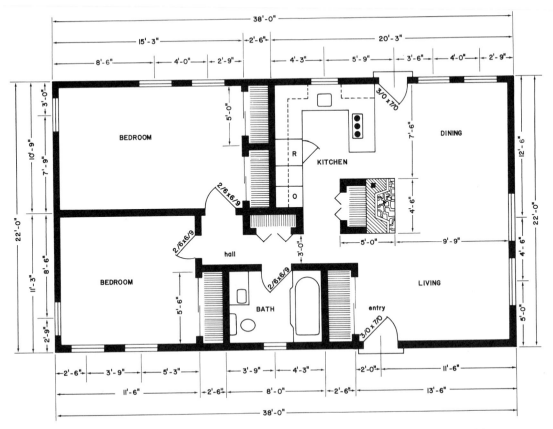

FIGURE 6-5 Residential Floor Plan (stud spacing 16 in. o.c. for both exterior walls and partitions; one bottom plate and two top plates throughout)

In this comparison of the two methods, the difference of 22 studs is about 10% on the low side for the approximate method. The variance will be greater or smaller depending on the layout of the walls and partitions for a particular floor plan. It is suggested that method 1 be used for greater accuracy in making stud take-offs.

Bracing. In order to give walls and partitions adequate stiffness and resistance to horizontal loads such as wind, bracing is installed at a 45-degree angle, or a full sheet of plywood is nailed to the studs at each corner or midpoint of a wall. Let-in bracing is usually a 1 × 4 which is notched into the studs as shown in Figure 6-4. Plywood bracing may be $\frac{3}{8}$- or $\frac{1}{2}$-in.-thick panels nailed to the studs as shown in Figure 6-6.

The amount and location of bracing will vary with the shape and size of the building. Two braces, or two panels of plywood, should be figured for the corners of

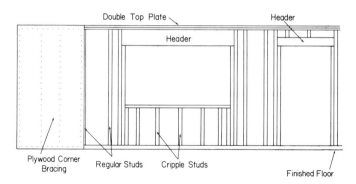

FIGURE 6-6   Typical Wall Framing

each main partition or exterior wall. Walls longer than 25 ft should have one or more additional braces or plywood panels between the wall corners.

Headers.   A wall header may be defined as a piece of dimension lumber spanning over an opening such as a door or window, supporting the load above that opening. A header is sometimes referred to as a "lintel." Figure 6-6 shows headers which are doubled over each opening, and they are at least 3 in. longer than the rough opening. The sizes of headers range from 2 × 4's to doubled 2 × 12's, depending on the span of the openings and the requirements of the local building code. Headers are also required across the openings in load-bearing partitions.

Firestops.   In balloon-type framing, firestops must be installed between studs on the second-story floor as shown in Figure 6-4. A firestop may be a 2 × 4 placed horizontally about halfway between the top and bottom plates in Western, or platform, framing. The purpose of the firestop is to prevent the creation of a draft, or flue effect, between studs. The firestop also adds strength to the wall.

When used as a stiffener or nailer, a horizontal member such as a 2 × 4 running between studs in a wall may be called a "girt." To estimate firestops or girts, use the plate-line length, in linear feet, of walls and partitions. From that value, subtract the linear feet of openings. Since this value is an approximation of the firestop required, it is not necessary to subtract stud thicknesses.

Ceilings

The construction of ceiling framing (Figure 6-7) is similar to the construction of floor system framing, except that header joists are not required, and the ceiling joists are usually smaller in size. Ceiling joists tie the opposite walls together, and they support the finished ceiling. The number of ceiling joists is determined in the same way that floor joists are determined. That is, the length of the run of the ceiling joists is divided

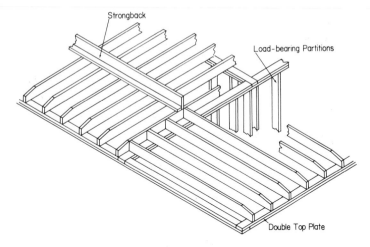

FIGURE 6-7   Ceiling Framing

by the spacing, or the length of the run is multiplied by the appropriate spacing factor. The size of the ceiling joist is determined by the length of the span, the spacing of the joists, and the grade of lumber. In order to coordinate with walls and permit the use of a wide range of finish materials, ceiling joists are usually spaced 12 or 16 in. on centers.

Walls or partitions that run parallel to ceiling joists must have a nailing strip on top of their plates so that the ceiling finish material can be fastened securely. When a partition runs parallel to the ceiling joists, it is called a "cross partition." A cross partition requires an extra ceiling joist, as shown in Figure 6-8.

To take-off ceiling joists spaced 16 in. on centers, the length of the supporting wall may be multiplied by $\frac{3}{4}$ and one joist added for a starter. The length of the ceiling joist is taken as an even number of feet from the working drawings. Although ceiling joists usually run across the narrow dimension of a building, sometimes they are placed to run at right angles to that direction.

When the spans of ceiling joists are excessive in length, stiffeners or "strongbacks" are placed at right angles to the joists at about midspan. Although the sizes of these stiffeners are determined by the design load, they often consist of 2×4's nailed flatways on the tops of the ceiling joists, and a 2×6 nailed vertically or edgeways to the side of the 2×4, as shown in Figure 6-7.

Roof System

A wood-framed roof system consists of a combination of various types of rafters, ridges, collar beams, gable studs, cut-in blocking, bracing, and sheathing. The basic parts of a framed roof system are shown in Figure 6-9.

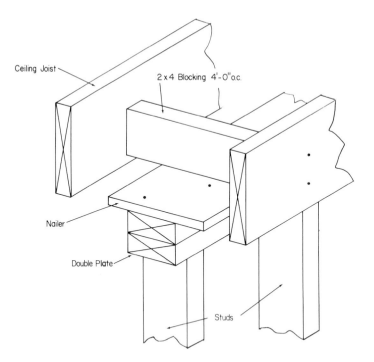

FIGURE 6-8   Nailing Strip on Top of "Cross Partition"

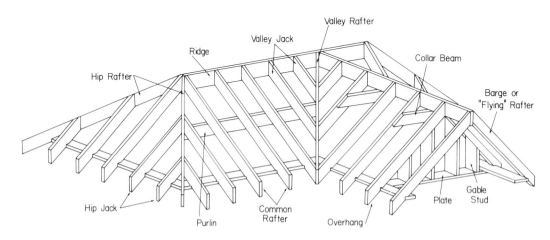

FIGURE 6-9   Roof System

Rafters.   The rafters in a gable-type roof are common rafters running the entire length of the plate line with barge or flying rafters of the same length forming the gable ends of the overhang. Common rafters are made from dimension lumber, and they are supported at the top by the ridge and at the bottom by the top plate of the exterior wall. The rafters in a hip roof are a combination of common rafters, hip rafters, valley rafters, hip-jack rafters, and valley-jack rafters.

To determine the number of pairs of common rafters required for a gable roof, divide the plate-line length by the rafter spacing (or multiply that length by the appropriate spacing factor), and add one extra pair for starters. The fact that a roof is hipped and that it has hip-jack and valley-jack rafters does not change the accuracy of this method for estimating purposes. That is because the part of the common rafter that drops off to make a jack rafter balances out to equal the length of a common rafter. For both the gable and hip types of roofs, the total number of common rafters is twice the number of pairs of rafters figured.

The next step is to figure the length, to the nearest even foot, required for a common rafter. This total will be the line length of the rafter, plus its tail length.

Although there are a number of ways to figure the length of a common rafter, two of the simplest ways involve using the Pythagorean theorem and the roof factor method.

The Pythagorean Theorem.   This theorem states that the square of the hypotenuse of a right triangle is equal to the sum of the squares of the base and altitude. Let the letter A stand for the hypotenuse, B stand for the base, and C stand for the altitude. The formula for the length of the hypotenuse becomes

$$A = \sqrt{B^2 + C^2}$$

The basic dimensions for a common rafter are shown in Figure 6-10. Note that the base of the right triangle is the "run" of the rafter, the altitude is the "rise," and the hypotenuse is the line length of the rafter. Therefore, using the Pythagorean theorem, the line length of a common rafter may be figured as follows:

$$\text{Line length of a common rafter} = \sqrt{\text{run}^2 + \text{rise}^2}$$

To determine the total length of the rafter, including the overhang, the horizontal projection must be added to the run. In addition, the vertical distance from the top of the plate to the horizontal line shown in Figure 6-10 must be added to the rise. This horizontal distance is the width of the soffit. The vertical distance is equal to the width, in feet, of the soffit times the slope of the roof. For example, if the soffit width is 2 ft 0 in. and the roof slope is 5 in 12, the vertical distance equals 2.0×5, or 10 in.

The Roof Factor Method.   A much simpler way to compute the length of common rafters is by using roof factors. A roof factor is a number that is multiplied times the run to give the length of the rafter. For example, using the right triangle for the Pythagorean theorem, and assigning values of 4 ft for the base, or run, and 3 ft for the altitude, or rise; the hypotenuse, or rafter length, equals 5 ft.

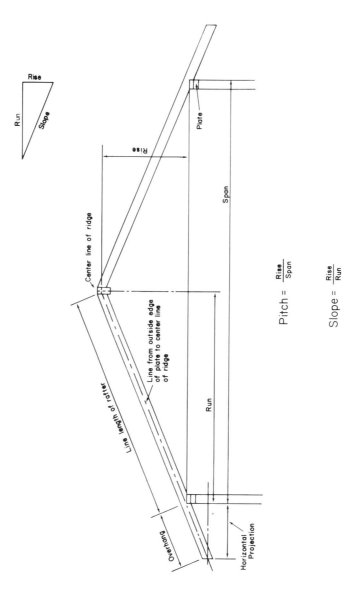

Pitch = $\dfrac{\text{Rise}}{\text{Span}}$

Slope = $\dfrac{\text{Rise}}{\text{Run}}$

FIGURE 6-10   Basic Dimensions for a Common Rafter

The roof factor concept is based on this relationship. The roof factor is the number by which 4 must be multiplied to get the number 5. In this case, that number is 1.25; and that number is the roof factor for a run of 4 and a rise of 3. Roof factors have been developed for various slopes using this principle, and they are included in Appendix C (Figure C-3).

A formula can be used to determine the roof factor for any slope. That formula is

$$\text{R.F.} = \sqrt{S^2 + 1}$$

where R.F. is the roof factor for a common rafter, and S is the slope of that rafter.

For example, to find the roof factor for a common rafter in a roof with a 9-in-12 slope, the procedure is as follows:

$$\text{R.F.} = \sqrt{\left(\frac{9}{12}\right)^2 + 1} = 1.25$$

To find the line length of a common rafter with a 9-in-12 slope and a 4-ft run, multiply 4 by 1.25, for an answer of 5 ft. If the run of the rafter, plus the horizontal projection, is 16 ft 4 in., the length of the rafter is 16.33×1.25, or 20.41 ft. The answer in feet and inches is 20 ft $4\frac{15}{16}$ in.

Roof factors may also be computed for hip and valley rafters using the following formula:

$$\text{R.F.(H\&V)} = \sqrt{S^2 + 2}$$

where R.F.(H&V) is the roof factor for hip and valley rafters, and S is the slope of the roof. To find the roof factor for a hip and valley rafter in a roof with a 9-in-12 slope:

$$\text{R.F.(H\&V)} = \sqrt{\left(\frac{9}{12}\right)^2 + 2} = 1.60$$

When the total length of the hip or valley rafter is to be figured, the horizontal projection of the overhang must be added to the run, and that total multiplied by the roof factor for that particular slope roof. For example, to determine the length of a hip or valley rafter in a roof with a 9-in-12 slope, a run of 14 ft 4 in., and a horizontal projection of 2 ft, the length is

$$16.33 \times 1.60 = 26.13 \text{ ft}$$

That answer, in feet and inches, equals 26 ft $1\frac{9}{16}$ in.

Ridges

The ridges of a roof system are usually one size larger than the rafters that frame into them. For example, a roof system with 2 × 6 common rafters will have 2 × 8's for ridge material. In a hip roof, the length of the ridge may be figured as the length at the eave line minus the width at the eave line, for a rectangular roof. In a gable roof of rectangular shape, the ridge is equal to the eave line.

Collar Beams

A collar beam, or collar tie, is a horizontal member that holds a pair of rafters together and prevents the rafters from spreading apart under load. Collar beams may be applied to every pair of rafters, or, depending on the load and slope of the roof, they may be installed on every other, or every third, pair of rafters. The number of collar beams may be figured by going back to the calculation for the number of pairs of rafters, and then multiplying that number by $1, \frac{1}{2},$ or $\frac{1}{3}$, depending on the spacing of the collar beams. The length of the collar beams may be estimated to be the length of the rafter run, plus whatever length it takes to reach the next highest even foot.

Gable Studs

Gable studs are usually 2×4's spaced 16 in. on centers and, if possible, placed to line up with the wall studs. The length of each gable stud may be figured as the distance between the ridge and the plate line, provided that only one-half of the building width is figured. By using this estimating technique, enough material will be ordered for the entire building width. The number of studs of this length is found by multiplying one-half the building width by $\frac{3}{4}$.

Bracing

To make an estimate of bracing required for roof framing, it is necessary to study the layout of the roof framing. Usually, a support is required at the midspan of each rafter, and this support is furnished by a member called a "purlin." To figure the length of purlins for a gable roof, multiply the length of the building by 2. To determine the number of braces supporting the purlins, multiply the length of the purlins by $\frac{1}{2}$ or $\frac{1}{3}$, depending on whether braces are installed every other, or every third pair of rafters. The lengths of the braces may be determined by scaling on a framing detail or cross section.

Roof Trusses

Roof trusses have an advantage over custom roof framing done at the jobsite because roof trusses may be prefabricated at ground level, and raised into position as a unit. This saves considerable labor costs. Depending on spans, loadings, and other factors, costs of roof trusses vary over a wide range.

Since the cost of material and labor can be figured reasonably close by the truss manufacturer, trusses are usually sold by the linear foot or by the unit. When the cost is so much per truss, the estimator determines the number of trusses required, and multiplies that number by the cost per unit.

FIGURE 6-11   Monopitch Roof Trusses

Figure 6-11 shows monopitch-type roof trusses on an apartment building. To estimate the number of trusses required, divide the plate-line length by the spacing of the trusses (or use the spacing constant), and add one for a starter. In the case of monopitch trusses, this result must be doubled.

Roof Sheathing

One of the easiest ways to estimate roof sheathing (Figure 6-12) is to use the roof factor for the slope of the roof for a particular building. To use this method, determine the horizontal projected area of the roof within the eaves and multiply that area by the roof factor. The final step is to divide that product by 32 when standard-size plywood roof sheathing is being used. A waste factor of 5 to 10% may be added depending on whether the roof is gable or hip, and job conditions.

For example, to determine the number of 4 ft × 8 ft plywood panels required to deck the roof of the house shown in Figure 6-13, the procedure is as follows:

1. Compute the horizontal projected area within the eave lines.

$$42 \times 56.58 = 2376.36 \text{ SF}$$

2. Multiply that figure by the roof factor for a 5-in-12 roof slope.

$$\text{R.F.} = \sqrt{\left(\frac{5}{12}\right)^2 + 1} = 1.0833 \times 2376.36 = 2574.31 \text{ SF}$$

FIGURE 6-12   Plywood Roof Sheathing over Trusses

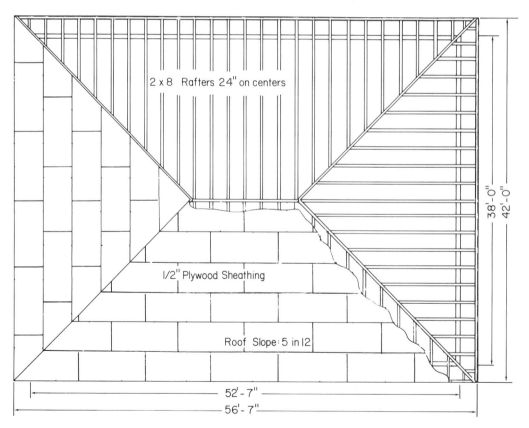

2 x 8 Rafters 24" on centers

1/2" Plywood Sheathing

Roof Slope: 5 in 12

38'-0"
42'-0"

52'- 7"
56'- 7"

FIGURE 6-13   Roof Framing and Sheathing

3. To get the number of pieces of plywood, divide by 32:

$$\frac{2574.31}{32} = 80.45$$

4. Add a 5% waste factor:

$$80.45 \times 1.05 = 84.47, \text{ or 85 pieces of plywood}$$

FINISH CARPENTRY

For convenience, a finish carpentry take-off may be divided into exterior finish and interior finish. Under exterior finish are various siding materials, frieze boards, fascia, and soffits. Included in the interior finish category are cabinets and millwork, moldings, paneling, and stairs. It should be noted that the finish carpentry take-off is usually relatively simple since much of the information needed has already been figured.

Exterior Finish

Siding. In making a siding take-off, it is sometimes possible to merely subtract the area of the brick veneer from the area of the wall sheathing, with minor adjustments, to get the siding area. On houses having all sides covered with siding, the area of the siding equals the area of the wall sheathing. The area of siding is figured by the "square," which is equal to 100 SF. When 4 ft × 8 ft, 4 ft × 10 ft, or 4 ft × 12 ft panels of plywood are used for siding, figure the gross area of the exterior walls without deducting the areas of the windows and doors. By including these openings, a waste factor will automatically be included in the take-off. When horizontal strip-type siding is used, it is more accurate to figure the net area of the wall surface, and then add a waste factor of 15 to 20%, depending on the type of strip siding being used, the experience of the crews installing the siding, and so on.

Frieze boards. A frieze board is a trim member of the cornice and covers the joint where the soffit, or plancier, and the wall meet as shown in Figure 6-14. The frieze board runs around the perimeter of the building, and its length is equal to the exterior plate-line length. In the take-off, the frieze board is usually measured in board feet when its size is 2 × 6, or greater. When its size is less than 2 × 6, it is usually measured in linear feet.

Fascia. The fascia is often referred to as the "eave board" because it runs the full eave-line length of the building. The finish fascia may be a 1 × 6 or a 1 × 8. It is nailed to the rough fascia, which is usually a 2 × 6 or a 2 × 8 nailed to the ends of the rafter tails. The take-off of finish fascia is in board feet after the total linear feet has been determined.

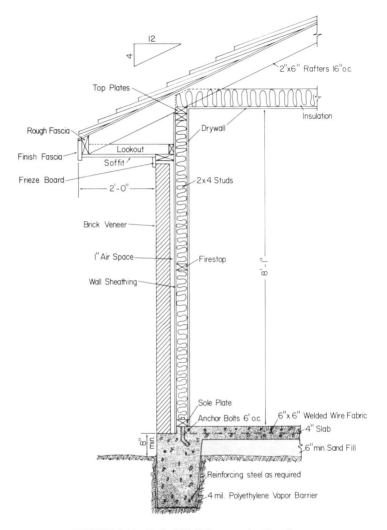

FIGURE 6-14   Typical Wall Construction Details

Soffits.   Soffit material must be of a material that is weather resistant. Exterior type plywood is a good example. The soffit take-off involves computing the square feet of material required and then converting that figure to the number of pieces of 4 ft × 8 ft panels of plywood required. An easy way to make a soffit take-off is to determine the stretch-out length (SOL) of the soffit, and multiply that length by the soffit width. For example, referring to Figure 6-13, note that the outside perimeter of the soffit in that figure is 197.16 ft, or 2 times the sum of 42 ft plus 56.58 ft. The stretch-out length is found using the formula

$$\text{SOL} = P_o - 4t$$

where P_o is the outside perimeter of the soffit, and t is the width of the soffit. Substituting in the formula gives

$$\text{SOL} = 197.16\text{ ft} - (4 \times 2) = 189.16\text{ ft}$$

To find the area of the soffit:

$$\text{Area} = 189.16 \times 2 = 378.32\text{ SF}$$

The final step is to figure the number of panels of plywood required, and to add a waste factor of 5%.

$$\text{Number of pieces} = \frac{378.32 \times 1.05}{32} = 12.41, \text{ or 13 pieces}$$

Interior Finish

Interior millwork in the average house includes doors, kitchen cabinets, vanities, shelving, and bookcases. Other houses or buildings may require, in addition, other millwork, such as stairs, fireplace mantels, ceiling beams, and wardrobes. In preparing an estimate for millwork items, allowance must be made for the labor required to assemble and install cabinets, shelving, stairs, and so on. For that reason, it is often advantageous to work out an agreement with the subcontractor concerning both fabrication and installation.

Cabinets.   To make a take-off for cabinets, list each cabinet separately giving the measurement from the wall, the height up the wall, and the length along the wall. With these data, it is possible to compute the tabletop area, or the area against the wall. There are a number of methods used to take-off cabinets. Some estimators price out cabinets by the front foot, or linear feet of length. Other estimators estimate cabinets and vanities on a unit price based on cubic content. For example, a kitchen cabinet may be listed as a base cabinet, 24 in. deep, 35 in. high, and 24 in. wide. Based on the cubic content, using those dimensions, the cabinet will cost so much per unit. To develop reliable data on cabinet take-offs, the estimator should check with local cabinet fabricators.

Moldings.   Moldings, such as aprons, base, casing, chair rails, closet poles, and so on, are taken off by the linear foot. Door trim, including headers, stops, and casing, are taken off by the individual opening.

Paneling.   The various types and patterns of paneling, including prefinished plywood and plastic-faced tempered hardboard, are figured by the square foot of wall area to be covered.

Stairs.   Folding stairs, or "disappearing stairs," installed in houses for access to the attic, are figured by the cost per unit, depending on the size and quality of the unit. Where it is necessary to custom-build stairs, details found in the working

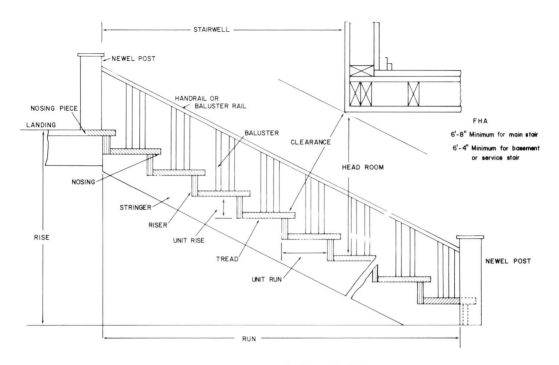

FIGURE 6-15   Parts of a Staircase

drawings furnish information for making a take-off. Figure 6-15 shows some of the parts of a staircase.

WORK EXERCISE 18: WOOD AND PLASTICS (CSI DIVISION 6)

Underpinning and Floor System

Objective: Using Figure 6-16, which shows the underpinning and floor system for a small structure, estimate the amount of dimension lumber and plywood required. Use a waste factor of 10% for lumber, and 5% for plywood.

Note
Figure band joists with the floor joists.

Compute the following:

1. Number and board feet of $2 \times 8 \times 20$ ft material required for girders
2. Number and board feet of $2 \times 8 \times 10$ ft material required for floor joists
3. Number and board feet of $2 \times 8 \times 12$ ft material required for floor joists
4. Number and board feet of $2 \times 6 \times 16$ ft material required for sills
5. Number and board feet of $2 \times 8 \times 20$ ft material required for headers
6. Number and square feet of pieces of $\frac{3}{4}$ in. $\times 4$ ft $\times 8$ ft plywood required

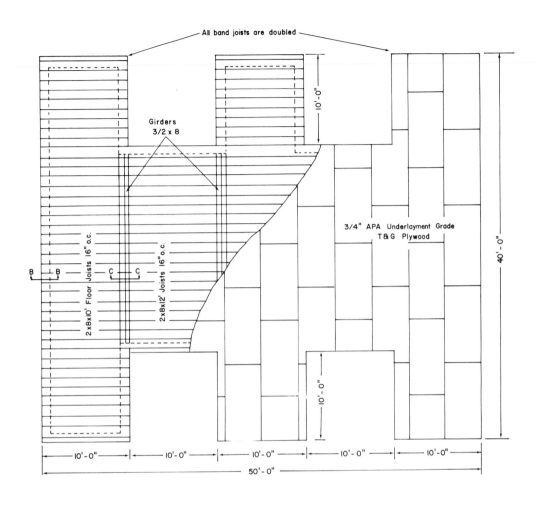

All band joists are doubled

Girders
3/2 x 8

3/4" APA Underlayment Grade
T&G Plywood

2x8x10' Floor Joists 16" o.c.

2x8x12' Joists 16" o.c.

B | B   C | C

40'-0"

10'-0"

10'-0"

10'-0"   10'-0"   10'-0"   10'-0"   10'-0"

50'-0"

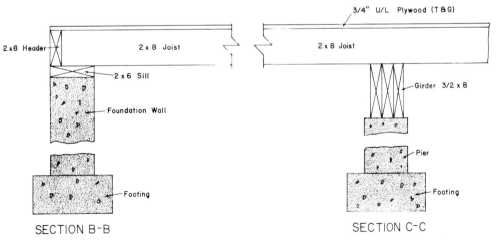

3/4" U/L Plywood (T&G)

2x8 Header

2 x 8 Joist

2 x 6 Sill

Foundation Wall

2 x 8 Joist

Girder 3/2 x 8

Pier

Footing

Footing

SECTION B-B

SECTION C-C

FIGURE 6-16 Floor Framing

111

WORK EXERCISE 19: WOOD AND PLASTICS (CSI DIVISION 6)

Floor System

Objective: Referring to Figures 6-17 and 6-18, estimate the amount of lumber and plywood required to frame the floor system for a small building. Use a waste factor of 10% for lumber and 5% for plywood.

Notes

1. The 2 × 12 band joists at the ends of the building are doubled, they are to be figured with the floor joists.
2. The partition that runs the length of the building in the basement is constructed with 2 × 6 studs. 2 × 6 girts, or blocking, run the full length of this partition.
3. Anchor bolts are spaced 4 ft on centers around the top of the foundation wall.

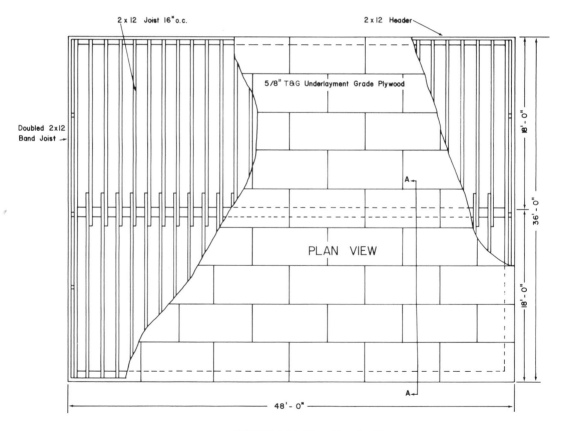

FIGURE 6-17 Floor Framing Plan

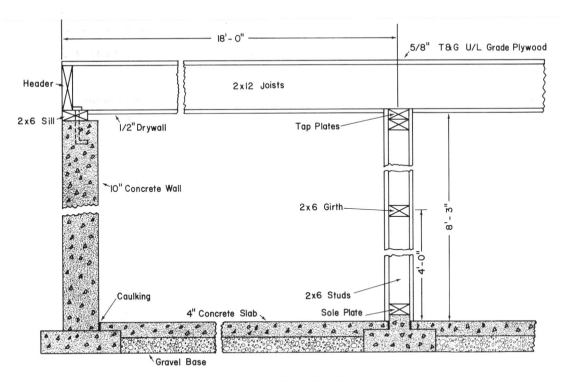

FIGURE 6-18   Typical Section

Compute the following:

1. Number and board feet of $2 \times 6 \times 16$ ft material required for sole plates and top plates
2. Number and board feet of $2 \times 6 \times 16$ ft material required for sills
3. Number and board feet of $2 \times 12 \times 20$ ft material required for floor joists
4. Number and board feet of $2 \times 12 \times 16$ ft material required for headers
5. Number and board feet of $2 \times 6 \times 8$ ft material required for studs
6. Number and board feet of $2 \times 6 \times 16$ ft material required for girts (blocking)
7. Number and square feet of pieces of $\frac{5}{8}$ in. $\times 4$ ft $\times 8$ ft plywood
8. Number of anchor bolts required

WORK EXERCISE 20: CONCRETE, AND WOOD AND PLASTICS (CSI DIVISIONS 3 and 6)

Foundation and Floor System

Objective: Using Figures 6-19 and 6-20, make a take-off of the material required for the foundation and floor system of a small building.

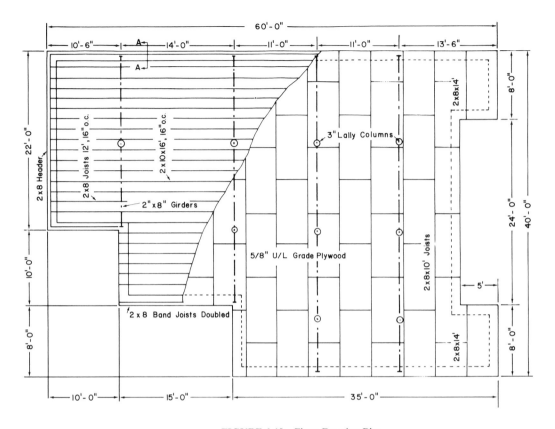

FIGURE 6-19   Floor Framing Plan

Notes

1. The girders are made up of three 2 × 8's, and their ends project into a 5-in.-deep pocket in the top of the basement wall.

2. In determining the weights of the reinforcement, use the formula: Weight of rebar per LF = $2.67D^2$.

3. Waste factors are 5% for concrete, 15% for sand cushion, 8% for reinforcement, 8% for lumber, and 5% for plywood.

Compute the following:

1. Number of cubic yards of concrete required for footings
2. Number of cubic yards of concrete required for basement wall
3. Number of cubic yards required for cushion sand
4. Number of cubic yards of concrete required for slab
5. Number of pounds of No. 5 rebar in footings
6. Number of pounds of No. 4 rebar in basement wall
7. Number and board feet of 2 × 10 × 16 ft material for sills

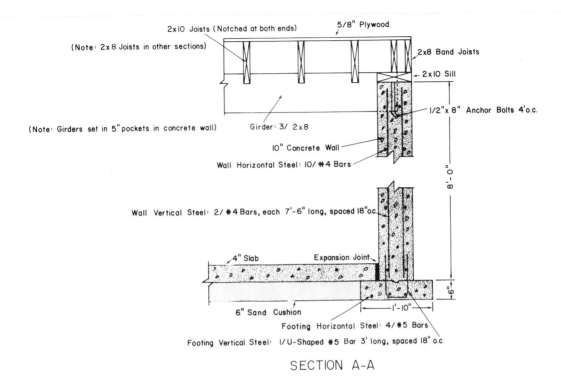

SECTION A-A

FIGURE 6-20 Typical Section

  8. Number and board feet of 2 × 8 × 16 ft of material required for girders
  9. Number and board feet of 2 × 8 × 16 ft material required for headers
 10. Number and board feet of 2 × 8 × 12 ft material required for floor joists
 11. Number and board feet of 2 × 8 × 14 ft material required for floor joists
 12. Number and board feet of 2 × 10 × 16 ft material required for floor joists
 13. Number and board feet of 2 × 8 × 10 ft material required for joists
 14. Number and square feet of pieces of $\frac{5}{8}$ in. × 4 ft × 8 ft plywood required
 15. Number of lally columns required
 16. Number of anchor bolts required

WORK EXERCISE 21: WOOD AND PLASTICS (CSI DIVISION 6)

Plates and Studs

Objective: Using the floor plan shown in Figure 6-21, make a take-off of the plates and studs required to rough-in the walls for a two-bedroom residence. In estimating the studs, use the two methods explained in Chapter 6.

Notes

1. Assume one sole plate and two top plates for both exterior walls and interior partitions.
2. Studs are spaced 16 in. on centers in both exterior walls and interior partitions.
3. In estimating sole plates, do not take out for door openings.
4. No waste factor is required for either method.

1. Determine the number and board feet of $2 \times 4 \times 16$ ft material required for plates.

Using method 1 for stud take-offs, determine the following:

2. Total length in LF of exterior walls plus interior partitions times $\frac{3}{4}$
3. Number of corners
4. Number of spaces between tees, plus number of spaces between tees and corners
5. Number of window and door openings times 2
6. Total number of studs required (obtained by adding items 2 through 5)

Using method 2 for stud take-offs, determine the following:

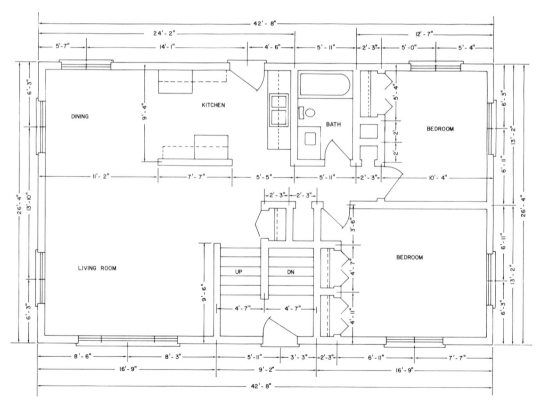

FIGURE 6-21 Floor Plan

7. Total number of studs required (obtained by multiplying the total length, LF, of exterior walls plus interior partitions, by 1)

WORK EXERCISE 22: WOOD AND PLASTICS (CSI DIVISION 6)

Roof Framing

Objective: Using Figure 6-22, make a take-off of the lumber and plywood required to frame the roof and ceiling for a small building.

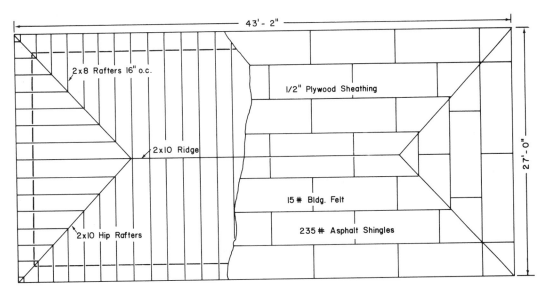

Roof Framing Plan

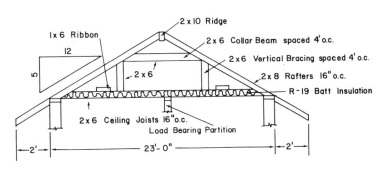

FIGURE 6-22   Roof Framing Plan and Section

Notes

1. The length of a common rafter may be found by multiplying the rafter's run by the roof factor for the roof slope.
2. The roof factor for a common rafter may be obtained from Appendix A, or it may be computed by using the formula

$$\text{R.F.} = \sqrt{S^2 + 1}$$

3. The length of a hip or valley rafter may be found by multiplying the common rafter's run by the hip or valley roof factor.
4. The roof factor for a hip or valley rafter may be obtained from Appendix A, or it may be computed by using the formula

$$\text{R.F.} = \sqrt{S^2 + 2}$$

5. Use a waste factor of 8% for lumber and 8% for plywood.

Compute the following:

1. Number and board feet of $2 \times 8 \times 16$ ft material for common and hip-jack rafters
2. Number and board feet of $2 \times 10 \times 20$ ft material required for hip rafters
3. Number and board feet of $2 \times 10 \times 18$ ft material for the ridge
4. Number and board feet of $2 \times 6 \times 8$ ft material for collar beams
5. Number and board feet of $2 \times 6 \times 10$ ft material for braces
6. Number and board feet of $1 \times 6 \times 12$ ft material for ribbons
7. Number and board feet of $2 \times 6 \times 12$ ft material for ceiling joists
8. Number and square feet of pieces of $\frac{1}{2}$ in. $\times$ 4 ft $\times$ 8 ft plywood required

WORK EXERCISE 23: WOOD AND PLASTICS (CSI DIVISION 6)

Roof Framing

Objective: Using Figure 6-23, make a take-off of the roof framing material required to frame the roof for a residence.

Notes

1. All rafters are spaced 16 in. on center.
2. The roof slope is 5 in 12.
3. Hip and valley rafters and ridges are one size larger than common rafters.
4. Use a waste factor of 5% for lumber and 8% for plywood.

Compute the following:

1. Roof factor for a common rafter with a 5-in-12 slope
2. Roof factor for a hip or valley rafter with a 5-in-12 slope

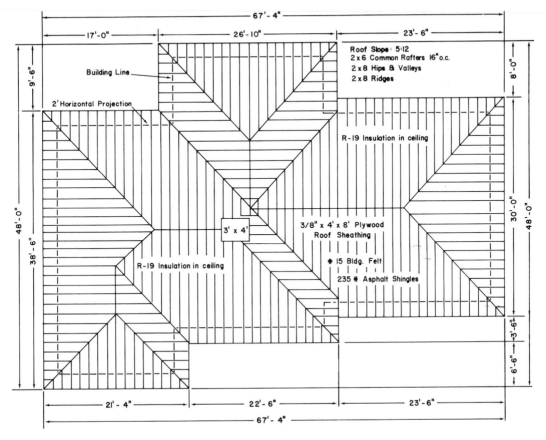

FIGURE 6-23   Roof Framing Plan

3. Number and board feet of 2 × 6 × 12 ft material required for common, hip-jack, and valley-jack rafters

4. Number and board feet of 2 × 6 × 16 ft material required for common, hip-jack, and valley-jack rafters

5. Number and board feet of 2 × 6 × 18 ft material required for common, hip-jack, and valley-jack rafters

6. Number and board feet of 2 × 8 × 16 ft material required for hip and valley rafters

7. Number and board feet of 2 × 8 × 20 ft material required for hip and valley rafters

8. Number and board feet of 2 × 8 × 24 ft material required for hip and valley rafters

9. Number and board feet of 2 × 8 × 16 ft material required for ridges

10. Number and square feet of pieces of $\frac{3}{8}$ in. × 4 ft × 8 ft plywood required for roof sheathing

WORK EXERCISE 24: WOOD AND PLASTICS (CSI DIVISION 6)

Exterior Finish

Objective: Using Figures 6-24 to 6-26, which show the elevations, floor plan, and section for a small house, make a take-off of the finish items required.

Notes

1. This house has a gable-type roof with a 4-in-12 slope.
2. "MDO" stands for medium-density overlaid plywood.
3. All plywood panels are 4 ft × 8 ft.
4. Figure a 1-in. lap for the horizontal strip siding.

Compute the following:

1. Number and square feet of pieces of $\frac{3}{8}$-in. MDO plywood for soffits
2. Number and square feet of pieces of $\frac{3}{8}$-in. MDO plywood required for gable ends
3. Number and board feet of 1 × 4 × 16 ft material required for frieze boards
4. Number and board feet of 1 × 6 × 16 ft material required for fascia
5. Number of pieces of $\frac{3}{8}$-in. 12 in. × 16 ft horizontal strip siding

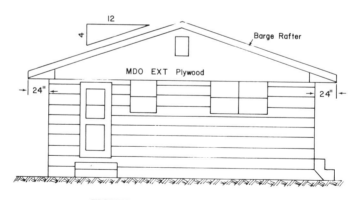

FIGURE 6-24 Residential Elevation

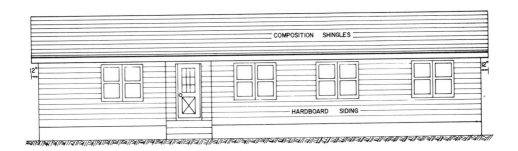

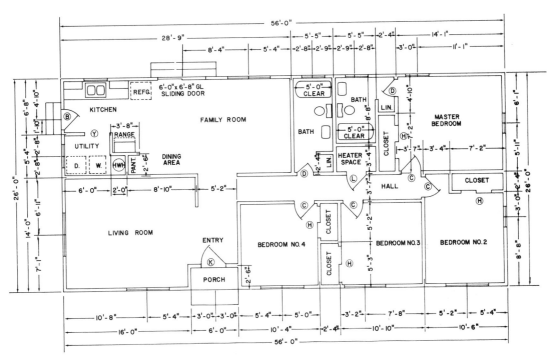

FIGURE 6-25   Floor Plan

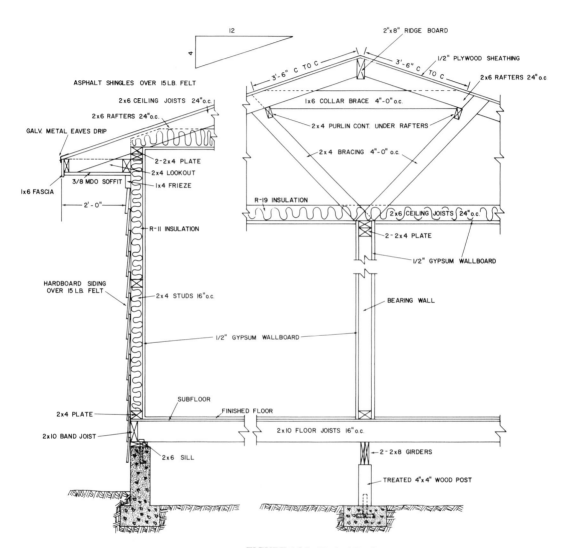

FIGURE 6-26 Typical Section

WORK EXERCISE 25: WOOD AND PLASTICS (CSI DIVISION 6)

Exterior Finish

Objective: Using Figures 6-27 and 6-28, which show the foundation plan and typical wall section for a small building, make a take-off of the exterior trim items listed below.

Notes
1. This building has a hip roof.
2. Use a waste factor of 5% for all items.
3. Plywood is 4 ft × 8 ft standard size.

Compute the following:

1. Number and square feet of pieces of $\frac{3}{8}$-in. plywood required for soffits
2. Number and board feet of 2 × 4 × 10 ft material required for lookouts

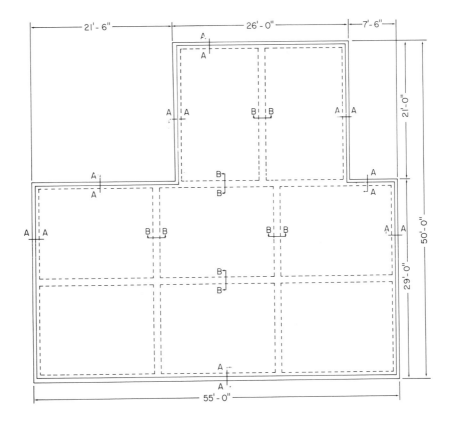

FIGURE 6-27   Foundation Plan

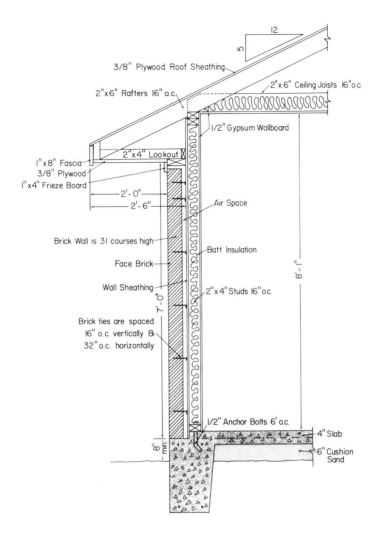

FIGURE 6-28   Typical Wall Section

3. Number and board feet of $1 \times 4 \times 16$ ft material required for frieze

4. Number and board feet of $1 \times 8 \times 16$ ft material required for fascia

7

Thermal and Moisture

Protection

OBJECTIVES

Upon completion of this chapter, the student will be able to:

- Estimate the amount of insulation required for a small building.
- Determine the number of squares of roofing material required for a small building.

GENERAL

In the process of considering ways to insulate and control moisture, the architect or designer must review many options. In the design and construction of a building envelope that will provide a satisfactory interior climate, the designer must specify the most efficient and economical materials. In addition, operating costs and maintenance factors must be considered. The building should be located on the site to take maximum advantage of natural light, solar heat, shade, and prevailing winds.

After the architect or designer has considered all of these factors and prepared a set of working drawings and specifications, the estimator must become familiar with those drawings and takeoff the items concerned with thermal and moisture protection. To do this, the estimator must become knowledgeable about the use of various materials, their covering capacities, their initial costs, and the cost of application. Included in the thermal and moisture protection of a building are waterproofing, dampproofing, insulation, roofing, flashing and sheet-metal work, and caulking.

WATERPROOFING

A number of methods, including integral waterproofing, plastic coating, membrane waterproofing, and the iron method, may be used to make a wall or surface water-resistant.

Integral Waterproofing

Water-repelling compounds may be mixed with cement, sand, and gravel while the concrete is being mixed. This method is called the "integral waterproofing method." When this method is used, the water-resisting compounds are usually estimated by the cubic foot or cubic yard of concrete to which the compound is added.

Plastic Coating

Another method, called the "plastic coating method," requires that various preparations be mixed with portland cement and sand, and applied as a plaster coat for waterproofing. When these coats are applied to concrete or masonry walls, the water-resisting additive is a liquid, paste, or powder mixed with one part portland cement and two parts sand. In estimating the cost, the thickness of the plaster coat determines the number of square feet of coverage of cement plaster coat for 1 CY of cement mortar. For example, for a coat of $\frac{1}{4}$ in. thickness, 1300 SF can be covered using 1 CY of cement mortar; $\frac{1}{2}$ in. thickness will cover 650 SF; $\frac{3}{4}$ in. thickness will cover 433 SF; and so on. Therefore, the take-off is in square feet or squares of wall area to be waterproofed. The unit cost is based on the thickness of plaster coat to be applied.

Membrane Waterproofing

Another method is the "membrane waterproofing method," which is constructed in place by building up asphalt- or tar-saturated plies that are applied with hot tar pitch or waterproofing asphalt. The membrane waterproofing method is probably the best method when it is applied correctly. This method blocks passage of moisture even when under hydrostatic pressure. The unit cost of this type of application will depend on the number of plies of saturated felt required. The quantity of pitch or asphalt required for mopping 100 SF, or 1 square, with one application is about 30 to 35 lb. Therefore, a two-ply application which requires three moppings will require from 50 to 105 lb of asphalt per square. A four-ply application which takes five moppings will require from 150 to 175 lb of asphalt per square of surface area to be covered. The makeup of the membrane will determine the cost per square. The take-off is simply a matter of figuring the total net area to be waterproofed. This area times the unit cost will give the cost of the waterproofing.

The Iron Method

If additional resistance to hydrostatic pressure is required, the "iron method" of waterproofing is recommended. This method uses fine metallic powder that is applied to either the inside or the outsides of the walls after the wall forms have been removed. In some cases, it is brushed on the tops of rough footings and floor slabs.

DAMPPROOFING

Materials Applied to Surfaces

Dampproofing differs from waterproofing in effectiveness. Dampproofing has less resistance to the passage of moisture under pressure. Where a hydrostatic pressure exists, the membrane method described previously should be used. However, dampproofing may be used to stop routine moisture. Dampproofing consists of applying one or more coats of heavy black paint to exterior concrete or masonry surfaces below grade. The number of coats and coverage will depend on the porosity of the surface of the wall. The take-off will be in square feet, or squares, of wall surface to be covered.

There are also colorless or transparent water-repellent treatments which may be applied to brick, stone, stucco, cement, and concrete surfaces. These treatments add to these materials' water resistance. Some of these treatments incorporate silicone-base products that make surfaces water repellent by penetrating deeply into the surfaces of the materials to which they are applied. The take-off for these treatments is in square feet, or squares, and the unit cost will be based on the method of application.

Vapor Barriers

Vapor barrier materials play an important role in controlling moisture movement. For example, a vapor barrier should be placed on the warm side of a wall, floor, ceiling, or roof to prevent moisture passing from the inside of a structure into the wall or ceiling cavities. Condensation occurs when high indoor humidities encounter low outside temperatures. This can be prevented through the use of a vapor barrier. Examples of vapor barrier materials are Sisalkraft, which is a reinforced building paper, and polyethylene, which is a plastic film of 0.004 or 0.006 in. thickness. Like other moisture protection materials, vapor barriers are figured using a unit cost per square.

INSULATION

Since the exterior and interior faces of a structure are usually dense materials and good conductors of heat, it is necessary to install insulating materials which are full of air pockets that will increase the insulating qualities. Insulating materials are rated for their thermal resistance which is measured by R values. The R value measures the temperature difference between two exposed faces required to cause 1 Btu to flow through 1 SF of the material per hour. In considering the R value of a wall system, individual R values can be added to one another to reach a total rating for the complete wall, floor, or ceiling construction. Some R values for typical building materials are shown in Figure 7-1.

The take-off for insulating materials is made by first determining the number of square feet, or squares, required for a particular type of insulating material. The next step is to apply the appropriate unit cost for that material.

Material	Thickness (in.)	R Value
Gypsum wallboard	$\frac{1}{4}$	0.22
Gypsum wallboard	$\frac{3}{8}$	0.32
Gypsum Type X	$\frac{1}{2}$	0.45
Plywood subflooring	$\frac{1}{2}$	0.62
Plywood subflooring	$\frac{5}{8}$	0.77
Plywood subflooring	$\frac{3}{4}$	0.93
Gypsum sheathing	$\frac{1}{2}$	0.45
Plywood siding	$\frac{3}{8}$	0.59
Wood drop siding	$\frac{3}{4}$	1.05
Stucco	1	0.20
Wood shingles	Exposure: $7\frac{1}{2}$	0.87
Face brick	Per in.	0.11
Common brick	Per in.	0.20
Concrete block	4	0.71
Concrete block	8	1.11
Cinder block	4	1.11
Cinder block	8	1.72
Cinder block	12	1.89
Concrete	4	0.32
Concrete	8	0.64
Lightweight concrete	2	2.22
Lightweight concrete	3	3.33
Asphalt shingles		0.41
Built-up roofing, 3-ply		0.78
Fiberglass batts	$3\frac{1}{2}$	11.00
Fiberglass batts	6	19.00
Fiberglass batts	$6\frac{1}{2}$	22.00
Fiberboard sheathing	$\frac{25}{32}$	2.06

FIGURE 7-1 R Values for Common Building Materials

ROOFING

Most roofing subcontractors bid on roofing, roof installation, and the sheet-metal work that goes with the roof, as a package. Figure 7-2 shows composition shingles being applied.

In making a roofing take-off, the estimator must compute the number of squares of surface area of the roof. The total square footage of the roof is determined, and then divided by 100 to obtain the number of squares required for that roof. The simplest and fastest way to determine the square footage of a roof is to multiply the horizontal projected area of the roof by the roof factor for the slope of that particular roof. In Chapter 6, the roof factor method for determining the length of common rafters was discussed. The formula for finding the roof factor for a roof was given as

$$\text{Roof factor} = \sqrt{S^2 + 1}$$

where S equals the slope of the roof. For example, the roof factor for a roof with a 4-in-12 slope is computed as follows:

$$\text{Roof factor} = \sqrt{\left(\frac{4}{12}\right)^2 + 1} = 1.054$$

The derivation of the roof factor formula, and the roof factors for slopes from 1 in 12 to 12 in 12, are given in Appendix C.

FIGURE 7-2   Composition Shingles over Plywood Sheathing

An example of estimating the number of squares of roofing for a small building may be shown by using the roof plan shown in Figure 7-3. To compute the number of squares of roofing required, first compute the square footage of the horizontal projected area of the roof, and then divide that value by 100. Next, multiply that value by the appropriate roof factor. Since this particular roof has a slope of 4 in 12, the roof factor is 1.054. The last step involves applying a waste factor of 5 to 10%, depending on job conditions and worker experience.

The procedure is as follows:

1. Find the horizontal projected area and divide by 100.

$$(28 \times 46) + (9 \times 27) = 1288 + 243 = 1531 \text{ SF}$$

$$\frac{1531}{100} = 15.31 \text{ squares}$$

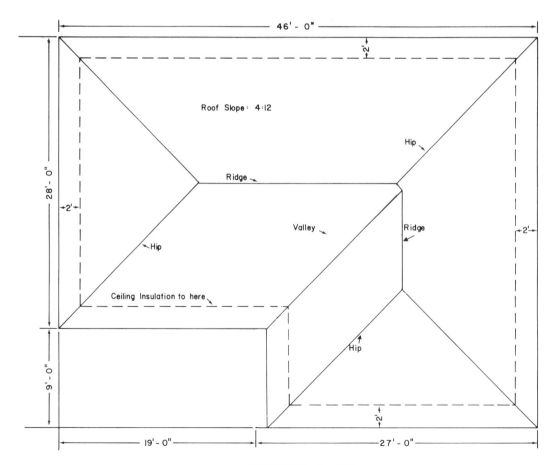

FIGURE 7-3   Roof Plan

2. Multiply the horizontal projected area by the roof factor.

$$15.31 \times 1.054 = 16.14 \text{ squares}$$

3. Apply a waste factor (say 5%).

$$16.14 \times 1.05 = 17 \text{ squares}$$

The roof of this building requires 17 squares of shingles.

Depending on the amount of exposure of shingles, the number of bundles will vary from 3 to 5 bundles per square. The total number of bundles required may be obtained by multiplying the number of squares of roofing required by the number of bundles per square. However, many estimators base their estimate on a base cost of so much per square, with that figure including labor costs, cost of so many bundles of shingles, cost of nails, cost of delivery, and cleanup.

FLASHING AND SHEET-METAL WORK

Sheet-metal flashing and counterflashing are available in galvanized metal, aluminum, and copper. Although flashing is usually measured by the linear foot, for widths 12 in. or greater, it may be measured in square feet. Metal valleys are frequently used with certain types of shingles and slate roofs. They are estimated by the linear foot with the width and kind of metal noted.

The cost of gutters or eave troughs will vary with the slope of the roof, the distance above ground, and the method used in hanging the gutters. The cost will be based on the total number of linear feet required.

Downspouts are available in both round and square shapes from $1\frac{1}{2}$- to 6-in. diameters. They extend from the gutters to the grade where an elbow diverts water away from the building.

CAULKING

In reducing or eliminating moisture penetration into a structure, caulking plays a very important role. Caulking should be used when two different materials meet, such as metal and brick, or wood and stone. Mortar joints between masonry and door and window frames must be filled solidly with an elastic caulking compound. Control joints and other wall openings should also be caulked. The cost of caulking is estimated at so much per linear foot, or so much per opening of a given size.

WORK EXERCISE 26: THERMAL AND MOISTURE PROTECTION (CSI DIVISION 7)

Insulation, Building Felt, and Asphalt Shingles

Objective: Using Figure 7-4, which shows a building with a hip roof, make a take-off of the insulation, building felt, and asphalt shingles required.

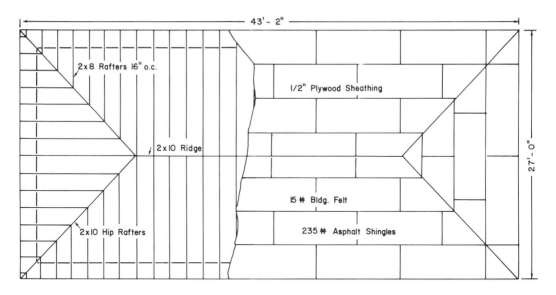

Roof Framing Plan

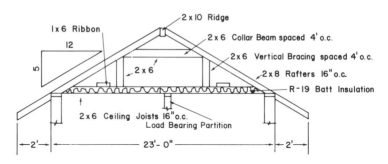

FIGURE 7-4   Roof Framing Plan and Section

Notes

1. Figure the number of fiberglass packages of insulation required in the ceiling only. Batts of 6 in. thickness, and an insulation value of *R*-19, are to be used.

2. One package of 6-in. fiberglass batt insulation is 15 in. × 39 ft, and will cover 49 SF.

3. One roll of 15-lb building felt will cover 400 SF.

4. Three bundles of 235-lb asphalt shingles will cover 1 square.

5. One bundle of ridge shingles will cover 88 LF.

6. The slope is 5 in 12 for this building, which has a hip roof.

7. No waste factors are required.

8. Figure hip shingles to run over the lengths of all hip junctions.

Compute the following:

1. Number of packages of 6-in. batt insulation
2. Number of rolls of building felt
3. Number of bundles of 235-lb asphalt shingles
4. Number of bundles of ridge and hip shingles

WORK EXERCISE 27: THERMAL AND MOISTURE PROTECTION (CSI DIVISION 7)

Ceiling Insulation, Building Felt, and Asphalt Shingles

Objective: Using Figure 7-5, which shows the roof framing plan of a building, make a take-off of the ceiling insulation, building felt, and asphalt shingles required.

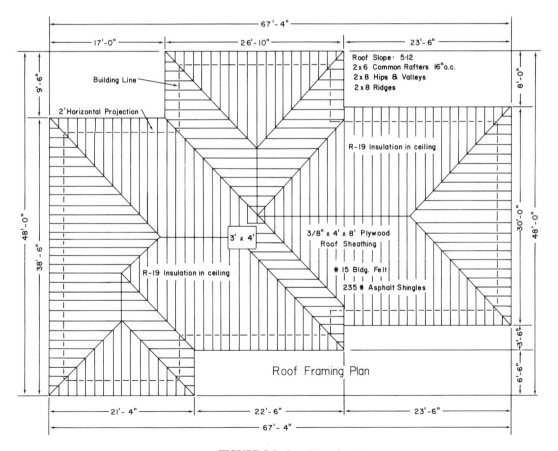

FIGURE 7-5 Roof Framing Plan

Notes

1. Figure the number of fiberglass packages of insulation required in the ceiling only. Batts of 6 in. thickness, and an insulation value of *R*-19, are to be used.

2. One package of 6-in. fiberglass batt insulation is 15 in. × 39 ft, and it will cover 49 SF.

3. One roll of 15-lb building felt will cover 400 SF.

4. Three bundles of 235-lb asphalt shingles will cover 1 square.

5. One bundle of ridge shingles will cover 88 LF.

6. The slope is 5 in 12 for this building, which has a hip roof.

7. No waste factors are required.

8. Figure hip shingles to run over the lengths of all hip junctions.

Compute the following:

1. Number of packages of 6-in. batt insulation
2. Number of rolls of building felt
3. Number of bundles of 235-lb asphalt shingles
4. Number of bundles of ridge and hip shingles

8

Doors and Windows

Upon completion of this chapter, the student will be able to:

- Identify the various types of doors and windows on a typical floor plan.
- Determine the number of doors and windows, by type, from a floor plan without the aid of door or window schedules.
- Determine the numbers of doors and windows, by type, using door and window schedules.

GENERAL

The procedure for taking off doors and windows from a set of working drawings is relatively simple when the architect or designer has provided detailed door and window schedules. For example, when a door schedule is provided similar to the one shown in Figure 8-1, the estimator is given the numbers of the various doors by types. With the quantities of doors known, it is just a matter of pricing out the units. Many factors are included in this process. It is important that the estimator make certain just what is included. For example, while most doors come as units with frames and hardware included, this is not always the case. Prehung metal doors are sometimes specified as exterior units on residential construction, and they may include frames, threshholds, weatherstripping, and a variety of locks. All residential working drawings do not include door and window schedules; or if they do, the quantities for each type of unit may not be included. In this event, the estimator must go to the floor plan

Symbol*	Quantity	Type	Rough Opening	Door Size	Manufacturer's Number
A	2	Flush	$3'\text{-}2\frac{1}{2}'' \times 6'\text{-}9\frac{1}{4}''$	$3'\text{-}0'' \times 6'\text{-}8''$	EF 36 B
B	6	Flush	$2'\text{-}10\frac{1}{2}'' \times 6'\text{-}9\frac{1}{4}''$	$2'\text{-}8'' \times 6'\text{-}8''$	IF 32 M
C	2	Flush	$2'\text{-}8\frac{1}{2}'' \times 6'\text{-}9\frac{1}{4}''$	$2'\text{-}6'' \times 6'\text{-}8''$	IF 30 M
D	8	Bifold	See manufacturer's specifications	$6'\text{-}0'' \times 6'\text{-}8''$	BF 36 AL
E	2	Sliding	$4'\text{-}2\frac{1}{2}'' \times 6'\text{-}9\frac{1}{4}''$	$4'\text{-}0'' \times 6'\text{-}8''$	IF 24 M
F	1	Garage	See manufacturer's specifications	$16'\text{-}0'' \times 7'\text{-}0''$	G 16 S

*Remarks

A. $1\frac{3}{4}''$ solid core, birch

B. $1\frac{3}{8}''$ hollow core, mahogany

C. $1\frac{3}{8}''$ hollow core, mahogany

D. Two units each 36″ wide, aluminum

E. $1\frac{1}{8}''$ hollow core, mahogany

F. Two-lite overhead sectional, aluminum

FIGURE 8-1   Typical Door Schedule

and elevations and determine the numbers of each type of unit. Therefore, it is essential that the estimator be familiar with the various types of doors and windows.

BASIC DOOR TYPES

Interior Doors

Doors may be classified as interior or exterior with a number of types within each class. Interior door types include flush, panel, bifold, sliding, pocket, accordian, and Dutch. Figure 8-2 shows these door types with their plan view symbols. Standard interior flush doors are smooth on both sides, hollow core, $1\frac{3}{8}$ in. thick, 6 ft 8 in. high, and come in a range of widths from 1 ft 6 in. to 3 ft 0 in. Figure 8-3 shows two interior flush doors.

Exterior Doors

Although exterior doors also come in flush types, they are usually solid core and thicker than interior doors. Exterior doors are $1\frac{3}{4}$ in. thick, 6 ft 8 in. high, and are usually of four basic types: flush, panel, sliding, or garage doors. While the first three types are usually 3 ft 0 in. in width, the garage doors are usually 8 ft 0 in. or 16 ft 0 in. in width, depending on whether they are single or double widths. The height of garage doors is usually 7 ft 0 in. Some common exterior door symbols are shown in Figure 8-4.

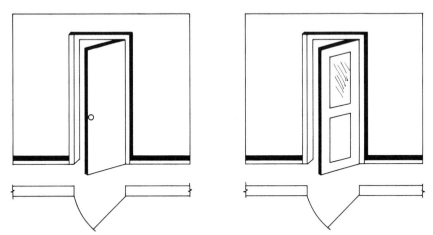

Flush and panel doors with plan view symbol

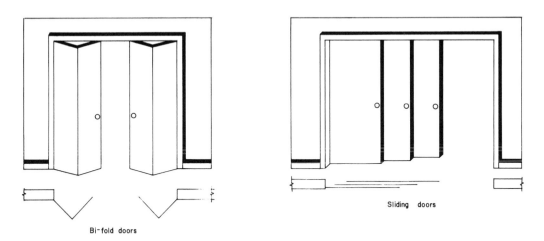

Bi-fold doors

Sliding doors

FIGURE 8-2 Types of Interior Doors

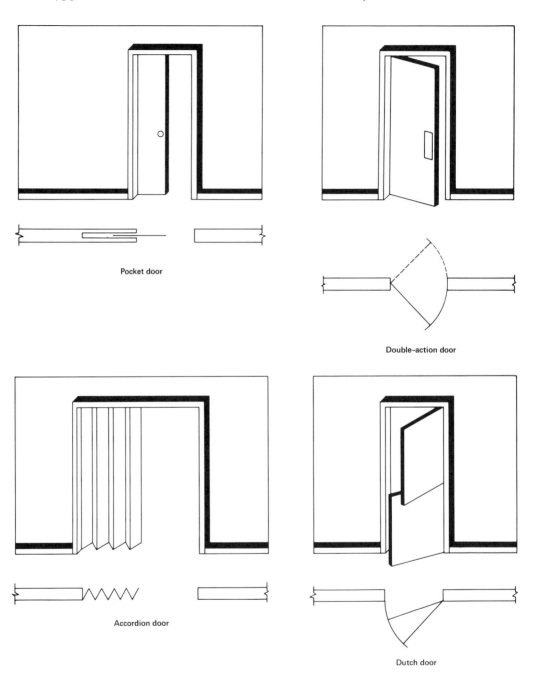

Pocket door

Double–action door

Accordion door

Dutch door

FIGURE 8-2 (cont.)

FIGURE 8-3   Interior Flush Doors

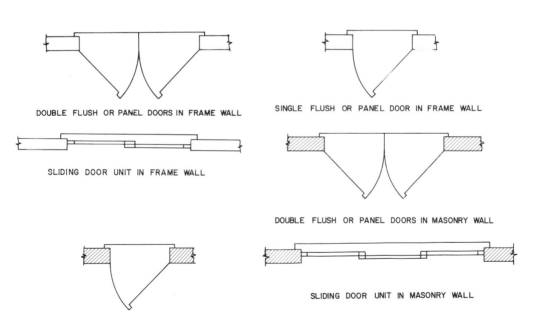

DOUBLE FLUSH OR PANEL DOORS IN FRAME WALL

SINGLE FLUSH OR PANEL DOOR IN FRAME WALL

SLIDING DOOR UNIT IN FRAME WALL

DOUBLE FLUSH OR PANEL DOORS IN MASONRY WALL

SINGLE FLUSH OR PANEL DOOR IN MASONRY WALL

SLIDING DOOR UNIT IN MASONRY WALL

FIGURE 8-4   Exterior Door Symbols

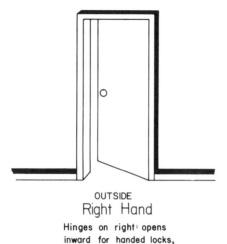

OUTSIDE
Right Hand

Hinges on right: opens
inward for handed locks,
specify RH.

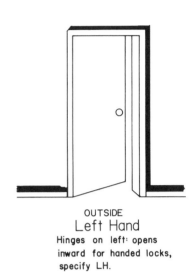

OUTSIDE
Left Hand

Hinges on left: opens
inward for handed locks,
specify LH.

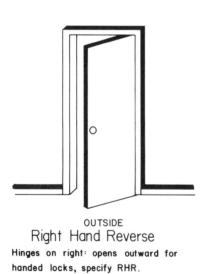

OUTSIDE
Right Hand Reverse

Hinges on right: opens outward for
handed locks, specify RHR.

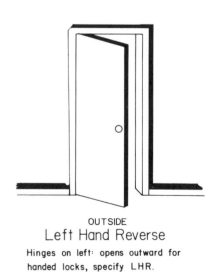

OUTSIDE
Left Hand Reverse

Hinges on left: opens outward for
handed locks, specify LHR.

FIGURE 8-5   The Hand of a Door for Locks

"Hand of the Door"

When ordering or specifying door locks, it is often necessary to describe the way in which the door swings. This may be done by facing the outside of the door, which is the street side of an entrance door, and the corridor side of an interior door. Figure 8-5 illustrates the standard procedure for determining the "hand of the door."

BASIC WINDOW TYPES

Windows serve many functions. In addition to admitting light and ventilation into the various rooms, they help create an atmosphere inside by framing an exterior view. They also add to the design of the exterior of the structure. Therefore, there are a number of types manufactured to meet these requirements in a variety of ways.

The four basic window types are double hung, casement, sliding, and awning. Other types include hopper, jalousie, bay, and picture windows. The symbols for these window types are shown in Figure 8-6. A double-hung window type is shown in Figure 8-7.

In making a take-off of windows, the procedure is similar to the one used in taking off doors. In the absence of a window schedule, the estimator must go to the floor plan and elevation to identify the numbers of each type of windows.

WORK EXERCISE 28: DOORS AND WINDOWS (CSI DIVISION 8)

Objective: Using Figure 8-8, which shows the floor plan of a four-bedroom residence, and using the door and window schedules, make a take-off of the doors and windows.

Notes
1. The term "mull" is defined as a window unit that consists of two single windows with a mullion between them.
2. The term "mullion" is defined as a fixed or movable post run vertically between lights in a window or door.

Determine the following:

1. Number of type B doors
2. Number of type C doors
3. Number of type D doors
4. Number of type H doors

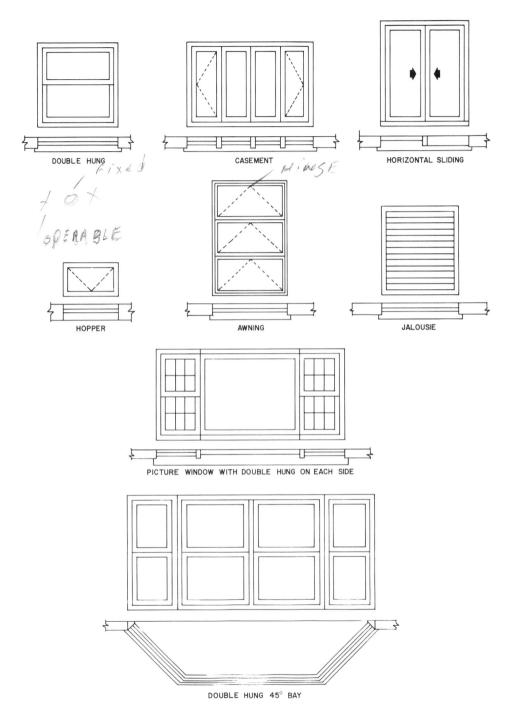

FIGURE 8-6 Types of Windows

FIGURE 8-7　Double-Hung Window

5. Number of type L doors
6. Number of type K doors
7. Number of type Y doors
8. Number of type Z doors
9. Number of type 2830 single windows
10. Number of type 2830 mull windows
11. Number of type 2840 single windows
12. Number of type 2840 mull windows

COMP DEDUCTS FROM
CALL OUT. OPENING

DOOR SCHEDULE		ROUGH OPENING	
SYMBOL	DESCRIPTION	HEIGHT	WIDTH
B	2'-8" x 6'-8" COMBINATION	6'-10 1/2"	2'-10 3/4"
C	2'-6" x 6'-8" INTERIOR FLUSH	6'-10 1/2"	2'-8 7/8"
D	2'-0" x 6'-8" INTERIOR FLUSH	6'-10 1/2"	2'-2 7/8"
H	4'-0" x 6'-8" BY-PASS	6'-10 1/2"	4'-1 7/8"
L	2'-6" x 6'-8" LOUVERED	6'-10 1/2"	2'-8 7/8"
K	3'-0" x 6'-8" CROSS BUCK	6'-10 1/2"	3'-2 3/4"
Y	3'-0" x 6'-10" OPENING	6'-10 1/2"	3'-0"

WINDOW SCHEDULE	ROUGH OPENING	
SYMBOL	HEIGHT	WIDTH
2830 SINGLE	37"	33"
2830 MULL	37"	65 1/8"
2840 SINGLE	49"	33"
2840 MULL	49"	65 1/8"

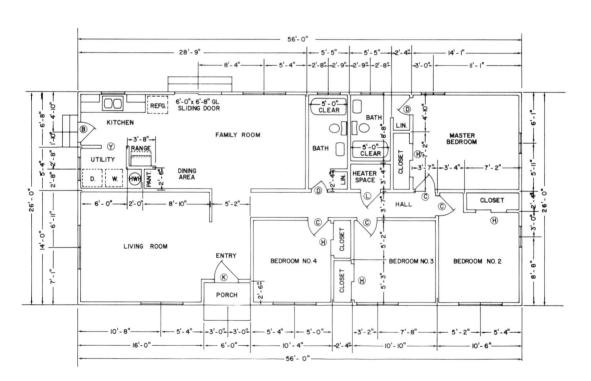

FIGURE 8-8   Floor Plan and Schedules

9

Finishes

OBJECTIVES

Upon completion of this chapter, the student will be able to:

- Take-off and estimate the quantity of gypsum drywall required for walls and ceilings of a residence.
- Take-off and estimate the quantity of paint required for painting the interior and exterior of a residence.
- Estimate the amount of carpet and resilient flooring required for a residence.
- Estimate the amount of ceiling tile required for the ceilings of the rooms of a residence.

GENERAL

Division 9 of the MASTERFORMAT lists the various materials needed to cover the framework of a structure. These items include lath and plaster, gypsum board, tile, terrazzo, acoustical materials, wood flooring, stone flooring, masonry flooring, resilient flooring, carpet, painting, and wall coverings. Although these items are normally done by subcontractors, the contractor must have a general knowledge of the materials and labor involved in many of these operations.

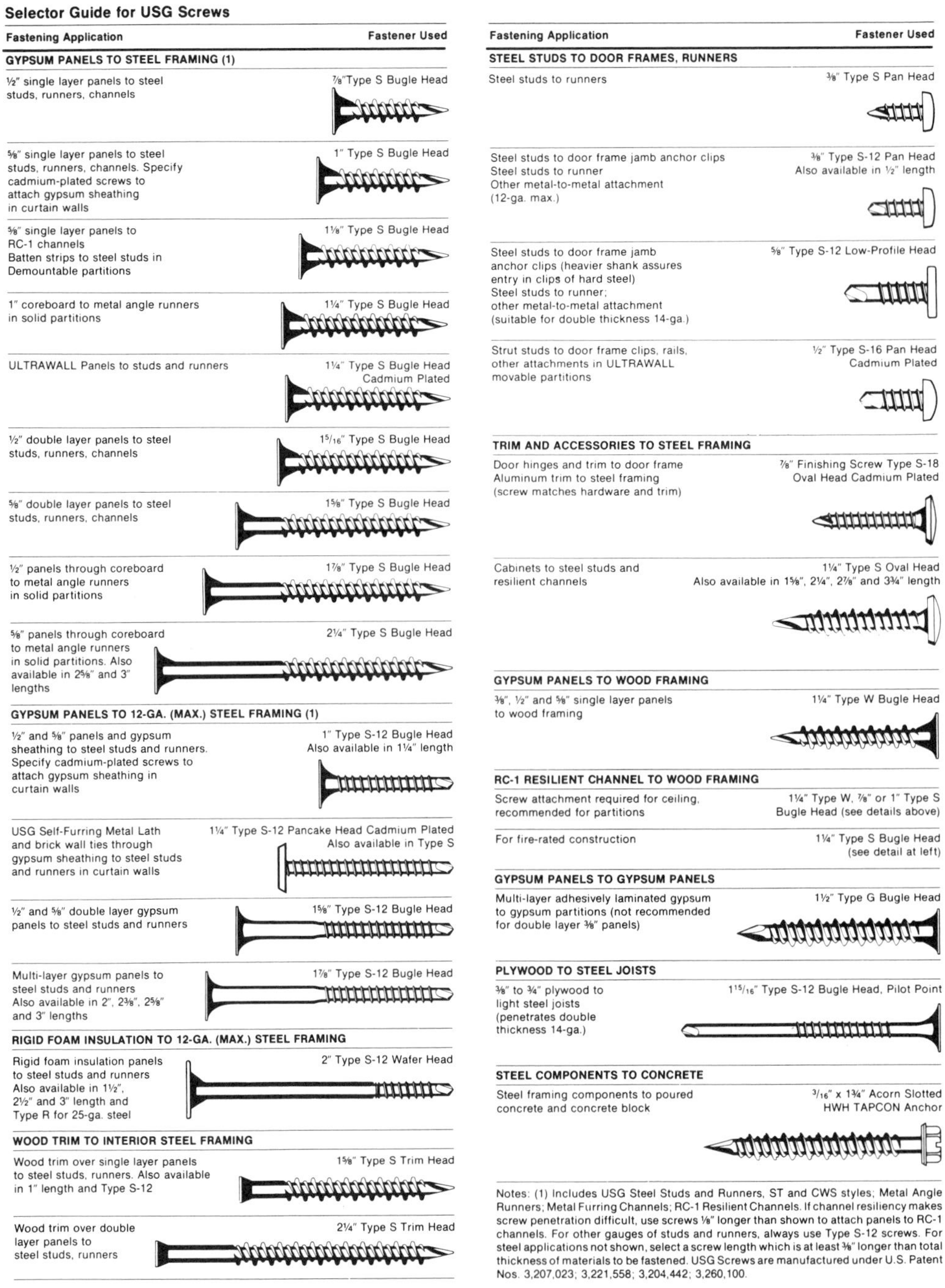

Selector Guide for USG Screws

Fastening Application	Fastener Used
GYPSUM PANELS TO STEEL FRAMING (1)	
½" single layer panels to steel studs, runners, channels	⅞" Type S Bugle Head
⅝" single layer panels to steel studs, runners, channels. Specify cadmium-plated screws to attach gypsum sheathing in curtain walls	1" Type S Bugle Head
⅝" single layer panels to RC-1 channels Batten strips to steel studs in Demountable partitions	1⅛" Type S Bugle Head
1" coreboard to metal angle runners in solid partitions	1¼" Type S Bugle Head
ULTRAWALL Panels to studs and runners	1¼" Type S Bugle Head Cadmium Plated
½" double layer panels to steel studs, runners, channels	1⁵/₁₆" Type S Bugle Head
⅝" double layer panels to steel studs, runners, channels	1⅝" Type S Bugle Head
½" panels through coreboard to metal angle runners in solid partitions	1⅞" Type S Bugle Head
⅝" panels through coreboard to metal angle runners in solid partitions. Also available in 2⅝" and 3" lengths	2¼" Type S Bugle Head
GYPSUM PANELS TO 12-GA. (MAX.) STEEL FRAMING (1)	
½" and ⅝" panels and gypsum sheathing to steel studs and runners. Specify cadmium-plated screws to attach gypsum sheathing in curtain walls	1" Type S-12 Bugle Head Also available in 1¼" length
USG Self-Furring Metal Lath and brick wall ties through gypsum sheathing to steel studs and runners in curtain walls	1¼" Type S-12 Pancake Head Cadmium Plated Also available in Type S
½" and ⅝" double layer gypsum panels to steel studs and runners	1⅝" Type S-12 Bugle Head
Multi-layer gypsum panels to steel studs and runners Also available in 2", 2⅜", 2⅝" and 3" lengths	1⅞" Type S-12 Bugle Head
RIGID FOAM INSULATION TO 12-GA. (MAX.) STEEL FRAMING	
Rigid foam insulation panels to steel studs and runners Also available in 1½", 2½" and 3" length and Type R for 25-ga. steel	2" Type S-12 Wafer Head
WOOD TRIM TO INTERIOR STEEL FRAMING	
Wood trim over single layer panels to steel studs, runners. Also available in 1" length and Type S-12	1⅝" Type S Trim Head
Wood trim over double layer panels to steel studs, runners	2¼" Type S Trim Head

Fastening Application	Fastener Used
STEEL STUDS TO DOOR FRAMES, RUNNERS	
Steel studs to runners	⅜" Type S Pan Head
Steel studs to door frame jamb anchor clips Steel studs to runner Other metal-to-metal attachment (12-ga. max.)	⅜" Type S-12 Pan Head Also available in ½" length
Steel studs to door frame jamb anchor clips (heavier shank assures entry in clips of hard steel) Steel studs to runner; other metal-to-metal attachment (suitable for double thickness 14-ga.)	⅝" Type S-12 Low-Profile Head
Strut studs to door frame clips, rails, other attachments in ULTRAWALL movable partitions	½" Type S-16 Pan Head Cadmium Plated
TRIM AND ACCESSORIES TO STEEL FRAMING	
Door hinges and trim to door frame Aluminum trim to steel framing (screw matches hardware and trim)	⅞" Finishing Screw Type S-18 Oval Head Cadmium Plated
Cabinets to steel studs and resilient channels	1¼" Type S Oval Head Also available in 1⅝", 2¼", 2⅞" and 3¾" length
GYPSUM PANELS TO WOOD FRAMING	
⅜", ½" and ⅝" single layer panels to wood framing	1¼" Type W Bugle Head
RC-1 RESILIENT CHANNEL TO WOOD FRAMING	
Screw attachment required for ceiling, recommended for partitions	1¼" Type W, ⅞" or 1" Type S Bugle Head (see details above)
For fire-rated construction	1¼" Type S Bugle Head (see detail at left)
GYPSUM PANELS TO GYPSUM PANELS	
Multi-layer adhesively laminated gypsum to gypsum partitions (not recommended for double layer ⅜" panels)	1½" Type G Bugle Head
PLYWOOD TO STEEL JOISTS	
⅜" to ¾" plywood to light steel joists (penetrates double thickness 14-ga.)	1¹⁵/₁₆" Type S-12 Bugle Head, Pilot Point
STEEL COMPONENTS TO CONCRETE	
Steel framing components to poured concrete and concrete block	³/₁₆" x 1¾" Acorn Slotted HWH TAPCON Anchor

Notes: (1) Includes USG Steel Studs and Runners, ST and CWS styles; Metal Angle Runners; Metal Furring Channels; RC-1 Resilient Channels. If channel resiliency makes screw penetration difficult, use screws ⅛" longer than shown to attach panels to RC-1 channels. For other gauges of studs and runners, always use Type S-12 screws. For steel applications not shown, select a screw length which is at least ⅜" longer than total thickness of materials to be fastened. USG Screws are manufactured under U.S. Patent Nos. 3,207,023; 3,221,558; 3,204,442; 3,260,100.

FIGURE 9-1 Fasteners for Gypsum Drywall. (Courtesy of the United States Gypsum Company.)

GYPSUM DRYWALL

Gypsum drywall, or gypsum board, is available in $\frac{1}{4}$, $\frac{3}{8}$, $\frac{1}{2}$, and $\frac{5}{8}$ in. thickness. The panel sizes come in 4-ft widths, and vary in length from 6 to 16 ft. Various metal accessories include corner beads and channels for edge trim around door and window openings.

Gypsum drywall may be fastened to the studs or ceiling joists with nails, screws, or adhesives. A number of different types of nails and screws are made for drywall applications. Figure 9-1 shows some of these fasteners.

Joint compound and perforated paper tape are used to cover and smooth joints for gypsum drywall applications. Figure 9-2 shows joints being smoothed over. Joint compound is available in 25-lb bags or ready mixed in 5-gallon cans. The perforated tape comes in 60-, 250-, and 500-ft rolls. The gypsum drywall can be finished with a texturing material which gives the appearance of textured plaster. Various patterns may be applied. These texturing materials may be applied with a brush, paint roller, or spray apparatus.

In making a take-off of gypsum drywall, the unit of measurement is the square foot. In estimating the areas of the walls and ceilings which are to receive gypsum drywall, it is necessary to estimate without taking out the areas of the doors and windows. Only large areas exceeding 30 SF should be subtracted from the gross wall area. Once the total net area of drywall is determined, the number of gypsum drywall panels is figured by dividing the net area by the surface area of one panel, or sheet. The result is then rounded off to the next full number.

FIGURE 9-2 Joint Treatment for Gypsum Wallboard

The quantity of nails, screws, tape, joint compound, and texturing material required is proportional to the area of the drywall to be installed, or to the number of sheets required. For example, based on the application of 1000 SF of gypsum drywall, about 6 lb of $1\frac{1}{4}$-in. annular ring nails, 50 lb of joint compound, 375 LF of perforated tape, and 20 to 50 lb of texturing compound will be required.

In estimating the cost of labor for gypsum drywall application, the estimator or contractor must draw on previous experience with similar types of jobs. Although the labor hours required to install 1000 SF of gypsum drywall will vary due to a multitude of factors, an approximation for labor can be given as shown in Figure 9-3.

Gypsum Drywall Application	Labor Hours per 1000 SF
Walls	18
Ceilings	24
Joints	9
Sprayed texturing	1–3

FIGURE 9-3   Gypsum Drywall Application

PAINTING

In the process of taking off quantities for exterior and interior painting, the actual surface to be painted is measured as accurately as possible using the working drawings. From the drawings, it is possible to estimate the quantity of materials required

FIGURE 9-4   Spray Painting

by determining the square footage, or number of squares required. Since the coverage will vary depending on the texture and porosity of the material to be painted, each material requiring a different type of finish should be taken off separately. The area of coverage for a gallon of paint will also vary with the type of paint and the method of application. In Figure 9-4, paint is being applied with a spray apparatus. More paint per square is required for roller or brush application.

When estimating the wall areas for painting, it is not necessary to take out areas of door and window openings because the saving in painted area is more than offset by the cost of trimming around these openings.

Exterior Walls

To estimate the amount of exterior paint required for a gable roof house, determine the wall area to be painted plus the area of the gable ends. For example, take a house 60 ft long, 45 ft wide, and 20 ft from the ground to the eaves. Assume that the gable on the 45-ft side is 15 ft high. The amount of paint required is figured as follows:

1. Total length of all sides = 60 ft + 45 ft + 60 ft + 45 ft = 210 ft.
2. The area of the walls = 210 ft × 20 ft = 4200 SF.
3. The area of two gable ends = 45 ft × 15 ft × $\frac{1}{2}$ × 2 = 675 SF.
4. Total area to be painted = 4200 + 675 = 4875 SF.
5. Gallons of paint required for one coat (assume a coverage of 300 SF/gal):

$$\frac{4875}{300} = 16.25 \text{ gallons}$$

Interior Walls and Ceiling

To estimate the amount of interior paint required, take the measurements of the rooms to be painted, and determine the square footage of their surface areas. For example, for a room measuring 15 ft × 24 ft, 360 SF must be covered with a ceiling paint. Dividing this figure by the coverage per gallon (assume 300 SF per gallon), the paint required is 1.2 gallons. For walls, window and door openings should not be subtracted from the surface area because of the extra labor involved in trimming around those openings. To determine the amount of paint required for the walls of the same room with a ceiling height of 8 ft, add the lengths of the walls and multiply by 8.

$$15 \text{ ft} + 24 \text{ ft} + 15 \text{ ft} + 24 \text{ ft} = 78 \times 8 = 624 \text{ SF}$$

Divide this figure by the coverage per gallon of paint:

$$\frac{624}{300} = 2.08 \text{ gallons}$$

This results in the amount of paint required to paint the walls of this room with one coat of paint.

Interior Trim

In taking off interior trim such as casings, base, and jambs, the unit of measurement is the linear foot. It is assumed that a linear foot of trim equals 1 SF because the time required to finish these narrow members is about the same for finishing 1 SF of surface. Since both sides of interior doors will be painted, the door area must be doubled.

Windows

Window sizes are usually figured as the overall sizes, including the trim around the window. This overall area determines the amount of material and labor required for painting.

Labor

Labor costs are based on the amount of time it takes to paint a given area with a specified kind of paint. Since there are different methods of application and classes of work, it is difficult to estimate labor costs for painting. Accessibility is another factor influencing labor costs. For example, it will cost more to paint a cornice that is two stories high than to paint a wall one story high.

Approximate labor costs for painting may be figured for various types of work by applying current hourly wages for painters to the labor estimates shown in Figure 9-5.

Type of Painting	Hours of Labor per Square per Coat of Paint
Exterior	
Wood siding	0.6
Doors and windows	0.7
Cornice and trim	0.8
Interior	
Walls and ceilings	0.8
Trim	0.6

FIGURE 9-5   Labor Estimates

CARPET AND PAD

In taking off for the carpet and pad, the unit of measurement is the square yard. By knowing the length and width of a room, the square footage is computed, and that figure divided by 9 to get square yards. Charts like that shown in Figure 9-6 give

ROOM WIDTH IN FEET

ROOM LENGTH IN FEET	8'	9'	10'	11'	12'	13'	14'	15'	16'	17'	18'	19'	20'
8'	7	8	9	10	11	11	12	13	14	15	16	17	18
9'	8	9	10	11	12	13	14	15	16	17	18	19	20
10'	9	10	11	12	13	14	16	17	18	19	20	21	22
11'	10	11	12	13	15	16	17	18	20	21	22	23	24
12'	11	12	13	15	16	17	19	20	21	23	24	25	27
13'	12	13	14	16	17	19	20	22	23	25	26	27	29
14'	12	14	16	17	19	20	22	23	25	26	28	30	31
15'	13	15	17	18	20	22	23	25	27	28	30	32	33
16'	14	16	18	20	21	23	25	27	28	30	32	34	36
17'	15	17	19	21	23	25	26	28	30	32	34	36	38
18'	16	18	20	22	24	26	28	30	32	34	36	38	40
19'	17	19	21	23	25	27	30	32	34	36	38	40	42
20'	18	20	22	24	27	29	31	33	36	37	40	42	44
21'	19	21	23	26	28	30	33	35	37	40	42	44	47
22'	20	22	24	27	29	32	34	37	39	42	44	46	49
23'	20	23	26	28	31	33	36	38	41	43	46	49	51
24'	21	24	27	29	32	35	37	40	43	45	48	51	53
25'	22	25	28	31	33	36	39	42	44	47	50	53	56
26'	23	26	29	32	35	38	40	43	46	49	52	55	58
27'	24	27	30	33	36	39	42	45	48	51	54	57	60
28'	25	28	31	34	37	40	44	47	50	53	56	59	62
29'	26	29	32	35	39	42	45	48	52	55	58	61	64
30'	27	30	33	37	40	43	47	50	53	57	60	63	67
31'	28	31	34	38	41	45	48	52	55	59	62	65	69
32'	28	32	36	39	43	46	50	53	57	60	64	68	71
33'	29	33	37	40	44	48	51	55	59	62	66	70	73

SQUARE YARDS

FIGURE 9-6 Floor Areas (SY) Based on Room Size. (Courtesy of Armstrong World Industries, Inc.)

square yards required based on room size. However, the estimator may obtain the result almost as fast by multiplying room length by room width, and dividing by 9. Although the square footage for the pad is the same as that for the carpet, these items should be listed separately for pricing.

RESILIENT FLOORING

Resilient flooring includes asphalt tile, linoleum, rubber, cork, vinyl tile, and vinyl sheet flooring. Each type has its own qualities and should be selected on these considerations rather than on the basis of cost alone.

In computing the square footage of resilient flooring, a certain percentage must be added for waste. This percentage will depend on the size of the area to be covered, how it is shaped, the amount of cutting involved, and workers' experience. An approximate waste factor for resilient flooring may vary from 4% for an area of 1000 SF to 14% for an area of 50 SF or less. The square-foot price of flooring depends on colors, type, and quantity involved.

CERAMIC WALL AND FLOOR TILE

Ceramic tile provides a durable and maintenance-free surface that is available in many shapes and sizes. In estimating ceramic tile, the unit measurement is the square foot for floor and wall areas. Trim and base pieces are estimated by the linear foot. In figuring wall areas, door and window areas should be deducted. Trim required to finish around doors and windows must be added as separate cost items. In addition to the cost of ceramic tile, accessory materials such as wire mesh, sand, cement, mixing, and placing should also be included.

ACOUSTICAL TILE

Acoustical tiles are applied to the ceilings and walls for the purpose of absorbing and deadening sound. They are made of various materials and come in a variety of sizes such as 12 in. × 12 in. and 12 in. × 24 in. Most manufacturers package ceiling tiles with either 40 or 64 12 in. × 12 in. tiles to a carton, or 20 to 32 12 in. × 24 in. tiles to a carton. Therefore, in estimating ceiling or acoustical tile, it is convenient to determine the number of tiles of a particular size for a given area. One way to do this is to use reference charts similar to Figure 9-7. Based on a given room size, this chart gives the number of pieces of tile, gallons of adhesive, and the amount of furring required.

Room Size§	No. Pcs.† 12″ × 24″ Tile	Gals. Adhesive	Amt. Furring (lin. ft.) 12″ o.c.	Room Size§	No. Pcs.† 12″ × 24″ Tile	Gals. Adhesive	Amt. Furring (lin. ft.) 12″ o.c.
7 × 8	32	1	64	11 × 13	78	3	156
7 × 9	36	2	72	11 × 14	84	4	168
7 × 10	40	2	80	11 × 15	90	4	180
7 × 11	44	2	88	11 × 16	96	4	192
7 × 12	48	2	96	11 × 17	102	4	204
7 × 13	52	2	104	11 × 18	108	4	216
7 × 14	56	3	112	11 × 19	114	5	228
7 × 15	60	3	120	11 × 20	120	5	240
8 × 8	32	2	72	11 × 21	126	5	252
8 × 9	36	2	81	11 × 22	132	5	264
8 × 10	40	2	90	12 × 12	72	3	156
8 × 11	44	2	99	12 × 13	78	4	169
8 × 12	48	3	108	12 × 14	84	4	182
8 × 13	52	3	117	12 × 15	90	4	195
8 × 14	56	3	126	12 × 16	96	4	208
8 × 15	60	3	135	12 × 17	102	4	221
8 × 16	64	3	144	12 × 18	108	4	234
9 × 9	45	2	90	12 × 19	114	4	247
9 × 10	50	2	100	12 × 20	120	4	260
9 × 11	55	3	110	12 × 21	126	5	273
9 × 12	60	3	120	12 × 22	132	5	286
9 × 13	65	3	130	12 × 23	138	5	299
9 × 14	70	3	140	12 × 24	144	6	312
9 × 15	75	3	150	14 × 14	98	4	210
9 × 16	80	3	160	14 × 16	112	5	240
9 × 17	85	3	170	14 × 18	126	5	270
9 × 18	90	4	180	14 × 20	140	6	300
10 × 10	50	2	110	14 × 22	154	6	330
10 × 11	55	3	121	14 × 24	168	7	360
10 × 12	60	3	132	14 × 26	182	8	390
10 × 13	65	3	143	14 × 28	196	8	420
10 × 14	70	3	154	16 × 20	160	7	340
10 × 15	75	3	165	16 × 24	192	8	408
10 × 16	80	4	176	16 × 28	224	9	476
10 × 17	85	4	187	16 × 32	256	11	544
10 × 18	90	4	198	18 × 20	180	8	380
10 × 19	95	4	209	18 × 25	225	9	475
10 × 20	100	4	220	20 × 30	300	12	630
11 × 11	66	3	132	20 × 40	400	16	840
11 × 12	72	3	144				

§If room measurements do not come out to the even foot, use the next largest room size shown.

†When 12″ × 12″ tile is used, double the quantity of pieces shown in the 12″ × 24″ column.

FIGURE 9-7 Estimating Ceiling Tile. (Courtesy of Armstrong World Industries, Inc.)

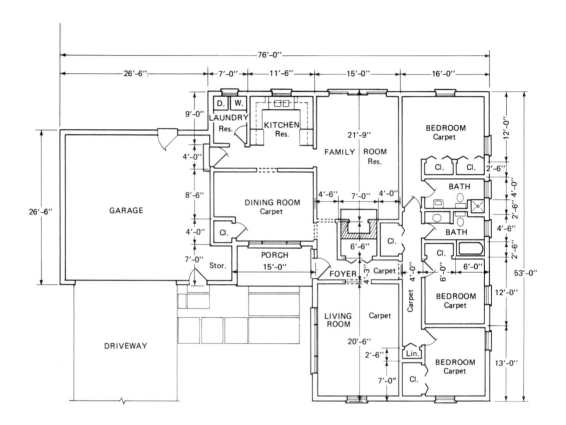

FIGURE 9-8   Floor Plan and Front Elevation for a Residence

WORK EXERCISE 29: FINISHES (CSI DIVISION 9)

Gypsum Drywall, Paint, Carpet, Resilient Flooring, and Ceiling Tile

Objective: Using the floor plan and front elevation for a residence, Figure 9-8, make a take-off of the gypsum drywall, interior paint, carpet and pad, and acoustical ceiling tile required.

Notes

1. Gypsum drywall is to be figured for the insides of all exterior walls, both sides of all interior partitions, and for all ceilings.
2. Figure gypsum panels to be 4 ft × 8 ft.
3. Gypsum drywall is to be applied to the inside walls and ceiling of garage.
4. In figuring gypsum drywall, do not take out for openings. No waste factor is to be added.
5. Estimate the number of gallons of flat interior paint required for interior walls of all rooms except the family room, which is to be paneled. Figure two coats and a coverage of 400 SF per gallon of paint.
6. Carpet and pad are to be laid in the living and dining rooms, the foyer, and in the three bedrooms.
7. Excepting the baths and the rooms receiving carpet, all other floor areas will be covered with resilient flooring. Use an 8% waste factor.
8. Ceiling tile is to be applied only to the family room and the bedrooms.

Compute the following:

1. Number of panels of 4 ft × 8 ft gypsum wallboard
2. Number of gallons of flat interior paint
3. Number of square yards of carpet material
4. Number of square yards of pad material
5. Number of square feet of resilient flooring
6. Number of square feet of acoustical ceiling tile

10

Mechanical

OBJECTIVES

Upon completion of this chapter, the student will be able to:

- Recognize the role of the mechanical subcontractor in the preparation of an estimate.
- Make an elementary take-off of some of the items involved in the heating and air conditioning of a small building.
- Make an elementary take-off of some of the items involved in plumbing a small building.

GENERAL

Under the heading "mechanical," the estimator must figure the work that includes heating, ventilation, air conditioning, and plumbing. This work is usually done under separate subcontracts with the various subcontractors making detailed estimates of the work involved. Although heating and plumbing are often given out by the owner to subcontractors separately from the general contractor, the general contractor is given jurisdiction over these subcontractors during the course of the job. On other occasions, the general contractor receives bids from the heating and plumbing sub-contractors and includes them in the final bid. In any event, the estimator should have a general knowledge of what is involved in a take-off for heating and plumbing. Since methods of handling excavation vary among the subcontractors, the estimator

156

must make certain of just what each subcontractor is responsible for, and make adjustments accordingly.

HEATING AND AIR CONDITIONING

Probably the most widely used heating method for residential work is the forced-warm-air method. This method uses blowers to circulate air in the system, and it is very efficient. In commercial work, various types of heating systems are used, including hot water boilers fueled by either oil or gas. Air conditioning and ventilating systems are usually included as a part of the heating estimate. In residential and small commercial work, the architect or designer usually specifies the type of system required, the capacity and performance expected. It then becomes necessary for the heating subcontractor to determine the details and design for a heating system to meet these requirements. When this is the case, it is very difficult for the general contractor to make an exact estimate for the heating and air conditioning for a particular job. Figure 10-1 shows outdoor condensing units on a split system furnishing air conditioning for two apartment units.

There are a number of approximate methods for determining heating and air conditioning costs. Some of these approximate methods involve developing unit costs per outlet in warm air heating, cost per radiator, cost per square foot of radiation for steam and hot water heating, cost per cubic foot of volume for space heating, or cost per square foot of living area. Since all of these methods give results that are only rough approximations of costs, the general contractor's estimator must rely on competent subcontractors for realistic, detailed, and dependent cost estimates.

FIGURE 10-1 Outdoor Condensing Units

PLUMBING

A plumbing estimate is usually broken down into rough and finish plumbing. Under rough plumbing, the following are included: sewer pipe, water and gas lines, vents and drains, and other items. Under the finish plumbing category are fixtures and their accessories.

On commercial work, a set of plumbing plans and specifications are prepared by an architect or engineer. These plans show the location, size, make, and type of the various fixtures and lines. From these plans, it is possible to determine the kinds, sizes, and material of pipes, stacks, traps, and so on.

Since residential plans seldom include detailed plumbing details and specifications, it is sometimes necessary for the subcontractor or the general contractor to work, and estimate, with insufficient information.

One rule of thumb for plumbing estimates is to figure that the plumbing will vary from about 3 to 12% of the cost of the project. This range is dependent on the number of fixtures and how they are arranged in the building. For an approximate cost for plumbing, a unit cost per fixture is sometimes used, plus costs for routing the supply and drainage lines.

The cost of sewer lines is estimated by determining the linear feet required for each size of line plus the number of fittings. Most plumbers estimate their sewer work at so much per linear foot, with this cost including excavation, material, equipment, and labor. Soil pipe, stacks, vents, and so on, are also estimated based on their lengths and the fittings required.

Black or galvanized piping may be estimated by length and size with a 60 to 75% additional cost for fittings. Various types of fixtures, such as sinks, lavatories, laundry trays, bathtubs, shower baths, hot water heaters, pumps, and so on, should be itemized and priced at current market prices.

The labor costs involved in roughing-in a one- or two-story house may be estimated as a percentage of the cost of the materials involved. This labor cost may range from 75 to 85% of the material cost.

Since the general contractor's estimator seldom makes a detailed estimate of all the plumbing required, the estimator's role is one of checking the plumbing subcontractor's bid. He or she must make certain that all of the items have been covered. The estimator must also make certain that the subcontractor's bid includes all fixtures and piping necessary to make the supply and drainage systems operable in accordance with the plans and specifications.

WORK EXERCISE 30: MECHANICAL (CSI DIVISION 15)

Heating, Ventilation, and Air Conditioning

Objective: Using Figure 10-2, which shows the floor plan for a small residence, make a takeoff of the ductwork items required for the heating and air conditioning system.

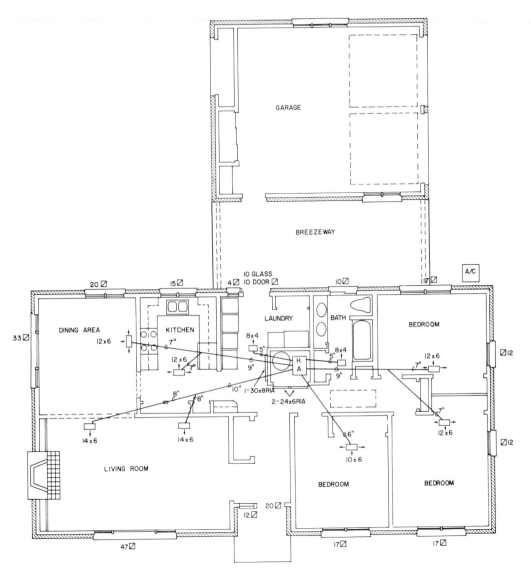

FIGURE 10-2   Heating, Ventilation, and Air Conditioning Plan

Notes

1. Study the floor plan and find the location of the furnace and air conditioning unit in the room adjacent to the laundry room.

2. Determine the number of registers, return-air grilles, supply heads, wyes, and elbows.

3. Estimate the lengths of each of the various diameters of ductwork. Since the ductwork distribution is run in the attic, figure additional ductwork length to run from the attic to the plenum of the furnace and air conditioning unit.

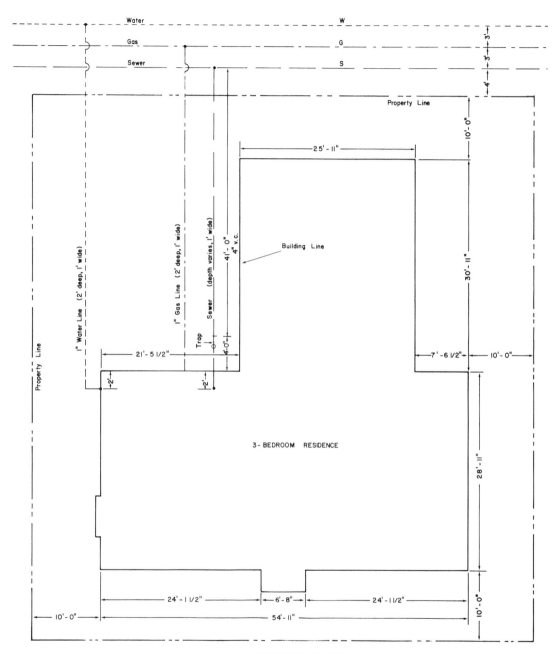

FIGURE 10-3 Utility Lines

4. The furnace/air conditioning unit is rated at 1250 cubic feet per minute (CFM) and includes filter, heating, cooling, coils, standard controls, 3 tons cooling, gas-fired.

WORK EXERCISE 31: MECHANICAL (CSI DIVISION 15)

Plumbing

Objective: Using Figures 10-3 and 10-4, make a take-off of the excavation required for the utility trenches, and some of the materials required for water, gas, and sewer lines for a small residence.

Notes

1. Plumbing materials for the interior of this residence are not required for this take-off.
2. Use the following waste factors:

> Excavation, 0%
> Cushion Sand, 14%
> Water, gas, and sewer pipes, 5%

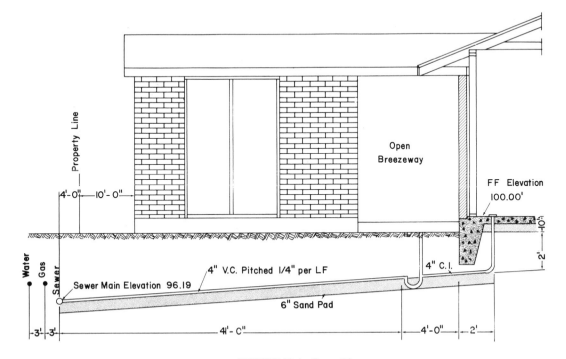

FIGURE 10-4 Sewer Line

Compute the following:

Excavation of utility trenches:

1. Cubic yards of excavation for water, gas, and sewer trenches

2. Cubic yards of cushion sand required for sewer line

Plumbing material:

3. Linear feet of 1-in. galvanized pipe for water line

4. Linear feet of 1-in. black iron pipe for gas line

5. Linear feet of 4-in. vitrified clay tile for sewer line

11

Electrical

OBJECTIVES

Upon completion of this chapter, the student will be able to:

- Recognize the role of the electrical subcontractor in the preparation of an electrical estimate.
- Make an elementary take-off of some of the items involved in the electrical requirements for a small building.

GENERAL

Depending on the magnitude of the lighting and services required, the electrical work in a modern residence may vary from 4 to 8% of the total cost of the house. To make an electrical quantity take-off, it is necessary to have, or to develop, a thorough knowledge of the information in the working drawings, the specifications, the building codes, and the regulations of the local electrical utility company.

Although a detailed electrical estimate is usually left to the electrical subcontractor, the general contractor's estimator can make some approximations concerning electrical costs. These approximations should be used as a guide in checking and reviewing the subcontractor's estimate.

One method used in making an approximate estimate of the electrical costs for a project is to base the estimate on the average cost per outlet. The total number of outlets is found by counting the convenience outlets in every room and location, all floor plugs, wall plugs, switches, and ceiling outlets. Some electrical symbols are shown in Figure 11-1.

Items that should be included in making a detailed electrical take-off are service-entrance equipment, branch circuits and wiring, convenience outlets and switches, and special-purpose items.

◯	CEILING OUTLET WITH INCANDESCENT FIXTURE	⊖•	SPLIT-WIRED DUPLEX RECEPTACLE-TOP HALF SWITCHED
◎	RECESSED OUTLET WITH INCANDESCENT FIXTURE	▥	FLOOR-MOUNTED RECEPTACLE
◯⊣	WALL-MOUNTED OUTLET WITH INCANDESCENT FIXTURE	T V	TV OUTLET MOUNTED 18" UP TO ₵ OF BOX
▭◯▭	CEILING OUTLET WITH FLUORESCENT FIXTURE	◤	TELEPHONE OUTLET
△◯△	FLOODLIGHT FIXTURE	▨	MAIN DISTRIBUTION PANEL
⊢◯⊣	FLUORESCENT STRIP	▬	LIGHTING-PANEL NUMERAL INDICATES TYPE
⊗	EXIT LIGHT, SURFACE OR PENDANT	⊬	BRANCH CIRCUIT CONCEALED IN CEILING OR WALLS SLASH MARKS INDICATE NUMBER OF CONDUCTORS IN RUN. TWO CONDUCTORS NOT NOTED
⊢⊗	EXIT LIGHT, WALL-MOUNTED	⊬	BRANCH CIRCUIT CONCEALED IN FLOOR OR CEILING BELOW
S	SINGLE-POLE SWITCH MOUNTED 50" UP TO ₵ OF BOX	⊬▲▲	INDICATES HOMERUN TO PANELBOARD; NUMBER OF ARROW HEADS INDICATES NUMBER OF CIRCUITS
S₃	THREE-WAY SWITCH MOUNTED 50" UP TO ₵ OF BOX	WP	WEATHERPROOF
S₄	FOUR-WAY SWITCH MOUNTED 50" UP TO ₵ OF BOX	Ⓙ	JUNCTION BOX
S₂	TWO-POLE SWITCH MOUNTED 50" UP TO ₵ OF BOX	D⊣	DIMMER CONTROL FOR LIGHTING FIXTURE
Sₗ	LOW VOLTAGE SWITCH TO RELAY	▭	CEILING ELECTRIC PANEL HEATER
Sᴅ	DOOR SWITCH	Ⓣ	DOUBLE-POLE THERMOSTATE FOR ELECTRIC HEAT
⊖	DUPLEX RECEPTACLE MOUNTED 18" UP TO CENTER OF BOX	Ⓓ	FIRE DETECTOR
⊘	DUPLEX RECEPTACLE MOUNTED 4" ABOVE COUNTERTOP	ⓈⒹ	SMOKE DETECTOR

FIGURE 11-1   Electrical Symbols

FIGURE 11-2   Service-Entrance Equipment

SERVICE-ENTRANCE EQUIPMENT

The service-entrance equipment consists of the service drop, service-entrance conductors, the meter, main control center, and the service ground. The main distribution box that receives the electricity and distributes it to various points in the house through branch circuits is called the service panel. Residential service-entrance equipment may be of the circuit breaker or switch and fuse type. Either type provides protection against overloads by opening the circuit if the current becomes too high. Figure 11-2 shows service-entrance equipment for a 16-unit apartment building.

Most residences have 240-volt service, which requires three wires. The service drop should be at least 10 feet above the ground, and at least 12 feet above driveways.

The main control center for residences should have a circuit-breaker-fuse panel of at least 100-ampere capacity. Circuit breakers will be priced according to their size and capacity.

BRANCH CIRCUITS

In the process of supplying electrical current to the appliances and outlets in a residence or building, it is necessary to group them together so that any overloads will cause only that particular circuit to go out. If the electrical system for a house consisted only of one giant circuit, the wire would have to be extremely large. In addition, if that circuit were to be overloaded, the entire house would be without power. The answer is to wire the house with a number of branch circuits. These circuits will have fuse sizes of 15- or 20-ampere capacity except for special need. One example is an electric range which will have a fuse size of 50- to 60-ampere capacity.

The *National Electrical Code®* specifies three types of branch circuits for residential use. They are (1) lighting circuits for lamps, radios, televisions, and similar 120-volt items; (2) special appliance circuits for all of the appliances in the kitchen, including toasters, mixers, blenders, coffee makers, and similar items; and (3) individual appliance circuits for permanently installed appliances such as washers, dryers, electric ranges, and similar items.

CONVENIENCE OUTLETS AND SWITCHES

The location of convenience outlets, switches, and light fixtures must be in accordance with electrical codes, furniture arrangements, function, and the personal preference of the owner or designer.

The *National Electrical Code®* requires a minimum of three convenience outlets in each room of a residence. Placement of convenience outlets along a wall should not be greater than 8 ft apart, and from 12 to 18 in. above the floor in most rooms. An exception is the kitchen, where most of the special appliance outlets are usually placed above the countertops. Most convenience outlets are the 120-volt duplex type, which have two receptacles.

It is recommended that at least one weatherproof outlet be placed on each exterior wall of a residence. In addition, it is a real advantage to have lighting fixtures and convenience outlets in the attic and crawl spaces of a house. The code requires that all convenience outlets must be grounded to prevent severe shocks to occupants and to minimize the danger of fire.

The location of switches in a residence should be about 48 in. above the floor. It is recommended that switches be located in convenient and visible places. As a safety precaution, bathroom switches must not be located within reach of a bathtub or shower.

Switches will vary in cost according to the type and quality. Most switches in a residence are the single-pole type. A single-pole type switch operates one fixture from one location. By using three-way switches, a fixture may be controlled from two locations. If a fixture is to be switched from three locations, two three-way switches and one four-way switch may be used. Another type of switch is the dimmer switch, which allows the intensity of light to be adjusted to the desired brightness.

SPECIAL-PURPOSE ITEMS

Among the special-purpose items requiring special outlets are television antenna outlets, burglar alarm systems, fire alarm systems, telephone jacks, intercom equipment, and others. Since these are specialized installations, they are usually installed by the manufacturer's representatives or the utility companies.

ESTIMATING ELECTRICAL COSTS

The electrical estimate is similar to the mechanical estimate in that the general contractor can only approximate the costs. For detailed costs, the general contractor must rely on experienced subcontractors. However, the general contractor and estimator must make certain that the electrical subcontractor covers all the electrical items in the working drawings and specifications.

WORK EXERCISE 32: ELECTRICAL (CSI DIVISION 16)

Objective: Using Figure 11-3, which shows a floor plan for a residence, make (a) a detailed electrical take-off, and (b) an approximate electrical take-off based only on the numbers of convenience outlets and switches required.

Notes

1. Study the floor plan and review the electrical symbols in Figure 11-1.

2. Take-off the electrical items and post these quantities to the appropriate blanks on Figure 11-4, the cost analysis form. This part of the exercise is the detailed electrical take-off.

3. For the approximate electrical take-off, compile the following data and make an approximation by assuming a cost of $30 per outlet/switch.

    a. Number of wall outlets

    b. Number of ceiling outlets

    c. Number of single-pole switches

    d. Number of three-way switches

    e. Total of items a through d

    f. Rough estimate of electrical cost (item e × $30)

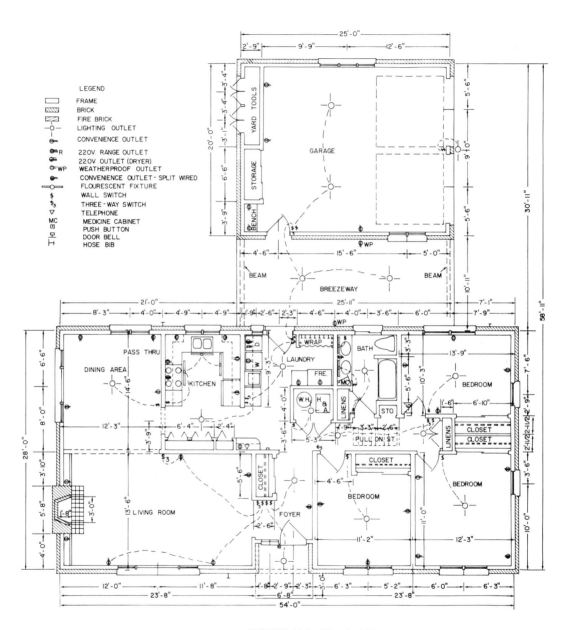

FIGURE 11-3   Electrical Plan

COST ANALYSIS

PROJECT __Three-Bedroom Residence_____   SHEET NO. _____
LOCATION _____   ESTIMATE NO. _____
ARCHITECT _____ OWNER _____   DATE _____
QUANTITIES BY _____ PRICED BY _____   CHECKED BY_____

DESCRIPTION	QUANTITY	UNIT	MATERIAL		LABOR		TOTAL COST	
			UNIT	TOTAL	UNIT	TOTAL	UNIT	TOTAL
CSI Div. 16 - Electrical								
1. Circuit Breaker Panel 225 amp., 3 pole, General purpose.		ea.						
2. Ceiling Outlets		ea.						
3. Outlet for garage floodlight		ea.						
4. Interior lighting fixtures: lamps, mounting harware and connections, incandescent		ea.						
5. Interior lighting fixtures: fluorescent		ea.						
6. Exterior lighting fixtures: with 100 watt lamps		ea.						
7. Duplex convenience outlets		ea.						
8. Triplex convenience outlets		ea.						
9. Waterproof convenience outlets		ea.						
10. Range outlet		ea.						
11. Dryer outlet		ea.						
12. Door chimes		ea.						
13. Single-pole switches		ea.						
14. 3-way switches		ea.						
15. Wiring, non-metalic cable (Romex), #14, 600V		LF						
16. Wiring, non-metalic cable (Romex), #12, 600V		LF						

FIGURE 11-4 Cost Analysis Form

12

Overhead and Profit

Upon completion of this chapter, the student will be able to:

- List the principal items involved in a contractor's overhead.
- Recognize the importance and critical nature of profit in the operation of a construction business.

GENERAL

Although overhead and profit have been left for the last chapter of this book on estimating, the critical importance of these items cannot be overemphasized. A contractor, or estimator, may prepare an accurate estimate of the costs to build a project, but without an understanding of the role that overhead and profit play, the chances of that firm staying in business are minimal. Too many contractors estimate and bid a job without adequate consideration of so-called "hidden costs" and a fair return for time and money expended. In addition to paying the firm a reasonable profit for work performed, the successful contractor must see to it that the firm also gets a fair return on the money invested in the business. Without that fair return, bankruptcy is just a question of time.

OVERHEAD

Costs that are not directly related to a specific construction activity or materials involved in doing a job are usually classified as "job overhead." Although some contractors look on these costs as hidden costs, they are not hidden, they are just uniden-

170

tified. Job overhead is the cost of various items that are incurred by just taking a job. These items include permits, supervisory salaries, fire insurance, field offices, bid bonds, warranties, and cleanup. Some of these are estimated item by item. Others are grouped together for a percentage markup. Regardless of how they are listed, they must be recognized as actual costs like labor and material. It is imperative that they be reflected in the final bid.

In many cases, the contractor's overhead is computed as a percentage of the cost of the labor and material required to do a job. However, some estimators prefer to use overhead percentages taken on just the labor costs. This method is used when the contractor furnishes the material, or when the costs of materials is extremely low relative to the cost of labor. A comparison of these two methods of applying overhead is shown in Figure 12-1.

METHOD 1 (Overhead applied to both labor and material).
Example for three jobs: (Assume an overhead of 20% and a profit of 10%).

	Job X	*Job Y*	*Job Z*
Labor	$50,000	$80,000	$20,000
Material	50,000	20,000	80,000
Sub Total	100,000	100,000	100,000
Overhead (20%)	20,000	20,000	20,000
Sub Total	120,000	120,000	120,000
Profit	12,000	12,000	12,000
Final Bid	$132,000	$132,000	$132,000

METHOD 2 (Overhead applied to labor only, but the overhead percentage is twice what it would be in Method 1 when both labor and material are figured).
Example for same three jobs used above: (Assume an overhead of 40% and a profit of 10%).

	Job X	*Job Y*	*Job Z*
Labor	$50,000	$80,000	$20,000
Material	50,000	20,000	80,000
Sub Total	100,000	100,000	100,000
Overhead (40%)	20,000	32,000	8,000
Sub Total	120,000	132,000	108,000
Profit (10%)	12,000	13,200	10,800
Final Bid	132,000	145,200	118,800

FIGURE 12-1   Comparisons of Two Methods for Applying Overhead Costs

General Contractor's Indirect Cost	Possible Markup of Both Material and Labor (%)
Field supervision	2.4
Main office expense	7.7
Worker's compensation and employer's liability	3.3
Field office	0.8
Performance bond	0.7
Unemployment tax	1.9
Social security	2.5
Tools and small equipment	0.4
Total	19.7

FIGURE 12-2   Overhead Costs

In those cases where the contractor's overhead is taken on both labor and material, the percentage markup may range from 10 to 25%. An example of a possible markup of 19.7% for job overhead is given in Figure 12-2. Keep in mind that this is just an example, and that the percentages will vary depending on many factors. Two of these factors are the size and organization of a construction company. Others include the ratio of supervision, job location, experience, and weather.

PROFIT

When all the bills have been paid and the contractor has been reimbursed for his or her investment, the remaining amount of the bid price, if any, is the profit. A more formal definition of profit is the excess of income for doing a job over labor costs, material costs, equipment costs, and overhead costs.

Normally, profit is computed as a percentage of the total labor, material, and equipment costs for a job, plus the overhead for that job. There is no pat answer as to what that percentage figure should be. Profit may range from zero to 15%, or higher. Some contractors set a minimum profit of 4% on the total bid, plus a 20% figure on net capital investment on that particular job.

On rare and unusual occasions, a contractor may take a job at zero profit to keep crews busy and continue to pay overhead costs. This does not necessarily mean that the contractor is losing money. He or she is recovering labor and material costs, and is being reimbursed for time and overhead.

However, it is obvious that a contractor cannot continue to take jobs at zero profit and expect to stay in business. The contractor must recognize that a reasonable profit is one of the primary reasons for doing a job. Profit, or lack of it, is the bottom line. The contractor must adjust the profit percentage to the risks involved, the amount of money invested in the job, the amount of projects under way in the area, and competitive conditions. A successful contractor must not only expend time and effort in doing a good-quality job; he or she must make money in the process.

WORK EXERCISE 33: OVERHEAD AND PROFIT

Objective: Using Figure 12-1, which shows a comparison of two methods of applying overhead, answer the multichoice statements below, and prepare a final bid based on the data given.

Notes

1. In method 1, an overhead percentage of 20% is applied to both labor and material.
2. In method 2, an overhead percentage of 40% is applied to labor only.
3. In both methods, profit percentage of 10% is assumed.

After reviewing the two methods shown in Figure 12-1, complete the statements below by selecting the best answer.

1. It is to a contractor's advantage to apply overhead to labor only (method 2) if the labor costs are:
   a. Greater than the material costs
   b. Less than the material costs
   c. The same as material costs
2. It is to a contractor's disadvantage to apply overhead to labor only if the material costs are:
   a. Greater than the labor costs
   b. Less than the labor costs
   c. The same as labor costs
3. As far as the final bid is concerned, there is no difference between the two methods as long as the labor and material costs are proportioned as they are in:
   a. Job X
   b. Job Y
   c. Job Z
4. The subtotal of $100,000 for each of the three jobs is a figure that represents:
   a. The total direct cost
   b. The total labor cost
   c. The total material cost
5. When the profit percentage is figured, it should be:
   a. Multiplied times the direct cost only
   b. Multiplied times the direct cost plus the overhead
   c. Multiplied times the overhead only

Prepare a final bid using the two methods described in Figure 12-1, and using the following new data:

1. A job is figured where the direct labor costs are $245,995.
2. The direct material costs are $327,840.
3. The overhead is to be figured at 19.7% for method 1, and 39.4% for method 2.
4. The profit markup, using both methods, is to be 9%.

	SUMMARY COST ESTIMATE				

Project _____ Estimate No. _____
Location _____ Sheet No. _____
Architect/Engineer _____ Date _____
Summary by _____

CSI Div. No.	Description	Estimated Material Cost	Estimated Labor Cost	Subcontracts	Total
1	General Requirements				
2	Site Work				
3	Concrete				
4	Masonry				
5	Metals				
6	Rough Carpentry				
	Finish Carpentry				
7	Thermal and Moisture Protection				
8	Doors				
	Windows				
9	Finishes				
10	Specialties				
11	Equipment				
12	Furnishings				
13	Special Construction				
14	Conveying Systems				
15	Heating, Ventilating, and Air Conditioning				
	Plumbing				
16	Electrical				
	Total Direct Cost				
	Overhead				
	Profit				
	Cost Per Square Foot _____			TOTAL BID	

FIGURE 12-3 Cost Estimate Summary Sheet

WORK EXERCISE 34: FINAL BID

Objective: Given estimates for various CSI Divisions as shown below, prepare a summary cost estimate sheet (Figure 12-3) for a residence.

Notes

1. Use an overhead percentage of 18.6% based on method 1, both labor and material.
2. Use a profit percentage of 11%.
3. Assume that the house for this takeoff has 1492 SF of living area.
4. Subcontractors have furnished estimates for masonry, mechanical, and electrical divisions.
5. Using the values given below, fill in the appropriate blanks on the summary cost estimate sheet (Figure 12-3).
6. Figure the percent of direct costs and percent of total bid for each of the separate divisions.
7. Give a total bid figure, and give the cost per square foot for this residence.

CSI Division	Description	Material	Labor	Subcontractors' Bids
2	Site work	$ 180	$ 452	
3	Concrete	4218	4112	
4	Masonry			$7288
5	Metals	426	417	
6	Rough carpentry	3842	3235	
6	Finish carpentry	2187	1197	
7	Thermal and moisture protection	1636	929	
8	Doors	1949	513	
8	Windows	1437	380	
9	Finishes	2814	3587	
15	Heating, ventilation, and air conditioning			3139
15	Plumbing			3338
16	Electrical			2347

Appendix A

The Stretch-Out-Length Concept

In the process of taking off quantities from a set of working drawings, the estimator spends a large amount of time in figuring areas and volumes. There are a number of useful techniques for reducing this expenditure of time. One of them is the stretch-out-length concept.

The stretch-out-length concept, or SOL, may be defined as the length of the centerline of the strips that form the perimeter of a figure such as the one shown in Figure A-1. For example, the 20 ft × 20 ft square in that figure has strips which are 2 ft thick. If the area of these 2-ft strips is to be computed, it may be done by figuring the areas of four rectangles, and then adding them for the total area. Two rectangles, measuring 2 ft × 20 ft, and two rectangles measuring 2 ft × 16 ft, give a combined area of 144 SF. A simpler way to figure this area is to determine the length of the centerline, and then multiply that length by the thickness of the strip. The formula for finding the stretch-out length, or SOL, is

$$SOL = P_o - 4t$$

where SOL is the stretch-out-length, P_o is the outside perimeter of the figure, and t is the thickness of the strip that runs around the perimeter of the rectangle. Substituting in the stretch-out-length formula, we have

$$SOL = 80 - (4 \times 2) = 72 \text{ LF}$$

With the stretch-out-length known, the area of the strip may be figured by multiplying that value by the thickness of the strip:

$$\text{Area} = SOL \times t \quad \text{or} \quad \text{Area} = 72 \times 2 = 144 \text{ SF}$$

For a simple square or rectangle, such as the one used in this example, it is relatively easy to figure the area of the strip by adding the areas of the four rectangu-

P_0 length around outside of figure

P_i = length around inside of figure

S.O.L. = length around centerline distance

S.O.L. = $P_0 - 4t$ where t = distance between P_0 and P_i, or width

$P_i = P_0 - 8t$

$P_0 = P_i + 8t$

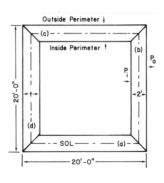

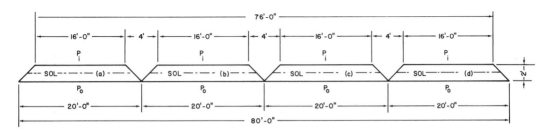

FIGURE A-1   Stretch-Out-Length Concept

lar strips. But the layouts encountered in actual buildings are more complex and rarely have only four corners. Some buildings may have as many as 12 or more corners, and a large number of individual rectangles would have to be considered. In those cases, the stretch-out-length concept can be a great timesaver. Regardless of the number of corners or insets involved in a layout, the basic stretch-out-length formula holds true. There is one qualification, and that is all corners must be 90-degree turns.

Figure A-2 shows a building with 12 corners and a 12-in.-thick wall. In order to figure the area of the top of that 12-in.-thick wall, the following steps may be taken:

1. Determine the length of the outside perimeter of the wall.
2. Compute the stretch-out length.
3. Compute the area of the top of the wall.

Step 1: An easy way to determine the outside perimeter of the wall is to adjust the perimeter to a rectangle, and then make changes as necessary. For example, in Figure A-2, the outside dimensions are 43 ft 0 in. by 46 ft 6 in., except for the fact that there are two setbacks, one measuring 5 ft 3 in. and one measuring 10 ft 2 in. In figuring the outside perimeter, P_o, add 43 ft 0 in. to 46 ft 6 in., and multiply the result by 2 to get 179 ft. But the two setbacks of 5 ft 3 in. and 10 ft 2 in. must be taken into account. Add them together and multiply by 2, for a result of 30 ft 10 in. When this

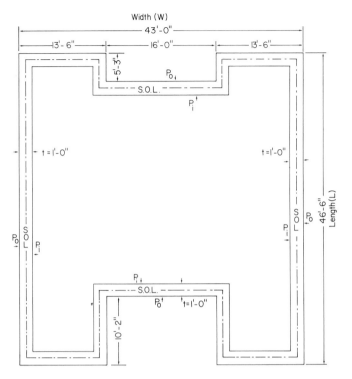

FIGURE A-2 Foundation Plan

sum is added to the 179 ft, the total of 209 ft 10 in. represents the adjusted perimeter length. Therefore, $P_o = 209.83$ ft.

Step 2: The stretch-out length is found by substituting in the formula as follows:

$$SOL = P_o - 4t$$
$$= 209.83 - (4 \times 1) = 205.83 \text{ ft}$$

Step 3: The area of the top of the strip is found by multiplying the stretch-out length times the thickness of the wall.

$$Area = SOL \times t$$
$$= 205.83 \times 1 = 205.83 \text{ ft}$$

If the volume of concrete required for the wall is required, the area of the top of the strip times the depth of the wall gives the answer in cubic feet. Assuming a depth of 8 ft for the wall, the volume is computed as follows:

$$Volume = \text{area of the top of the wall} \times \text{depth of the wall}$$
$$= 205.83 \times 8 = 1646.64 \text{ CF} = 61 \text{ CY}$$

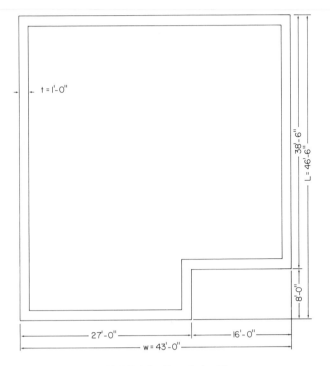

FIGURE A-3 Foundation Plan

Another common example is found in foundations with corner offsets such as the one in Figure A-3. This example differs from the foundation that has setbacks. It does not require an adjustment to the outside perimeter, which may be figured as follows:

$$P_o = 2 \times \text{width of building} + \text{length of building}$$

$$= 2(43.0 + 46.5) = 179.0 \text{ ft}$$

The stretch-out length for this example is

$$\text{SOL} = P_o - 4t = 179.0 - (4 \times 1) = 175 \text{ ft}$$

To summarize, when a foundation has setbacks, multiply the depth of each setback by 2 and add that value to the perimeter figured using the enclosing rectangle. When there are corner offsets, no correction is needed. The perimeter of the enclosing rectangle may be used without adjustment. Combine these rules when both types of variation from a rectangle are encountered in a foundation.

Appendix B

Formula for Determining

Weight of Reinforcing Steel

Reinforcing steel, or rebar, is estimated by the pound or by the ton. In converting reinforcing steel from linear feet to pounds, it is necessary to multiply the length to be converted by the weight per linear foot for that particular size rebar. Although the weight-per-foot value for standard reinforcing steel bars is given in various references (Figure 3-2), the estimator may compute this value by using a simple formula. This formula is

$$\text{Weight per foot of rebar} = 2.67D^2$$

where D is the diameter of the rebar in inches. For example, the weight per foot for a No. 3 rebar is:

$$\text{Weight per foot of No. 3 rebar} = 2.67(0.375)^2 = 0.376 \text{ lb/ft}$$

The formula is derived as follows:

1. Assume a weight for reinforcing steel of 490 PCF.
2. The cross-sectional area of a round rebar $= 0.7854D^2$.
3. The volume of a rebar 1 ft long equals: $12 \times 0.7854D^2$, with the result in cubic inches. Divide this value by 1728 to convert to cubic feet.
4. Therefore, the weight per foot of any round rebar equals

$$\frac{12 \times 0.7854D^2 \times 490}{1728} = 2.67D^2$$

Appendix C

Roof Factor Formulas for Common Rafters, and Hip and Valley Rafters

1. *Objective:* Derive a formula for a factor which, when multiplied by the run of a common rafter, will give the length of a common rafter (Figure C-1).
Prove: Rafter length = roof factor × run, or

$$c = \text{R.F.} \times a \qquad (1)$$

$$\text{R.F.} = \frac{c}{a} \qquad (2)$$

In Figure C-1,

$$c = \sqrt{a^2 + b^2} \qquad (3)$$

Substituting for c in Equation (2) gives

$$\text{R.F.} = \frac{\sqrt{a^2 + b^2}}{a} \qquad (4)$$

$$= \sqrt{\frac{a^2 + b^2}{a^2}} \qquad (5)$$

$$= \sqrt{1 + \frac{b^2}{a^2}} \qquad (6)$$

$$= \sqrt{\left(\frac{b}{a}\right)^2 + 1} \qquad (7)$$

Since the slope, S, of a roof is defined as the rise divided by the run, substitute S for b/a and the formula becomes

$$\text{R.F.} = \sqrt{S^2 + 1} \qquad (8)$$

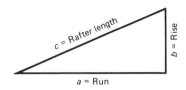

FIGURE C-1   Rise, Run, and Common Rafter Length

Example: To find the roof factor for a common rafter with a slope of 5 in 12:

$$\text{R.F.} = \sqrt{\left(\frac{5}{12}\right)^2 + 1} = 1.0833$$

This factor, 1.0833, multiplied by the run of the rafter will give the length of the rafter.

2. *Objective:* Derive a formula for a factor which, when multiplied by the run of a common rafter, will give the length of a hip or valley rafter (Figure C-2).
   Prove: Rafter length $=$ roof factor $\times$ run, or

$$e = \text{R.F.} \times a \tag{1}$$

$$\text{R.F.} = \frac{e}{a} \tag{2}$$

In Figure C-2,

$$e = \sqrt{b^2 + d^2} \tag{3}$$

Substituting for e in Equation (2) gives us

$$\text{R.F.} = \frac{\sqrt{b^2 + d^2}}{a} \tag{4}$$

In Figure C-2,

$$d^2 = a^2 + a^2 = 2a^2 \tag{5}$$

Substituting for d^2 in Equation (4), we have

$$\text{R.F.} = \frac{\sqrt{b^2 + 2a^2}}{a} \tag{6}$$

$$= \sqrt{\frac{b^2 + 2a^2}{a^2}} \tag{7}$$

$$= \sqrt{\frac{b^2}{a^2} + 2} \tag{8}$$

$$= \sqrt{\left(\frac{b}{a}\right)^2 + 2} \tag{9}$$

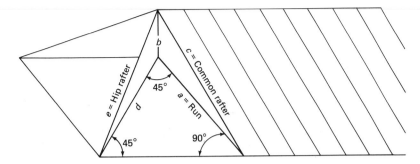

FIGURE C-2   Rise, Run, and Hip Rafter Length

Since the slope, S, of a roof is defined as the rise divided by the run, substitute, S for b/a and the formula becomes

$$\text{R.F.} = \sqrt{S^2 + 2} \tag{10}$$

Example: To find the roof factor for a hip or valley rafter in a roof with a 5-in-12 slope:

$$\text{R.F.} = \sqrt{\left(\frac{5}{12}\right)^2 + 2} = 1.4743$$

This factor, 1.4743, multiplied by the run of a common rafter will give the length of the hip or valley rafter.

3. *Objective*: Given the slope and total run of a rafter, use a table of roof rafter factors to calculate the lengths of the common and hip rafters in a roof system.

Roof Slope	Common Rafter Factor	Hip or Valley Rafter Factor
2 in 12	1.0138	1.4240
3 in 12	1.0308	1.4362
4 in 12	1.0541	1.4530
5 in 12	1.0833	1.4743
6 in 12	1.1180	1.5000
7 in 12	1.1577	1.5298
8 in 12	1.2018	1.5634
9 in 12	1.2500	1.6008
10 in 12	1.3017	1.6415
11 in 12	1.3566	1.6853
12 in 12	1.4142	1.7320

FIGURE C-3   Roof Rafter Factors

Procedure: To calculate the length of a rafter, determine the width of the structure (outside to outside of wall plates), divide by 2, add the horizontal overhang, and then multiply this value by the appropriate roof factor (Figure C-3).

Example: Determine the common and hip rafter lengths for a house that is 25 ft wide, with an overhang of 3 ft, and a roof slope of 4 in 12.

Common rafter length $= (12.5 + 3.0)(1.0541) = 16.3$ ft

Hip rafter length $= (12.5 + 3.0)(1.4530) = 22.5$ ft

Appendix D

Masterformat: Master List of Section Titles and Numbers

(Courtesy of the Constructions Specifications Institute.)

BIDDING REQUIREMENTS, CONTRACT FORMS*, AND
CONDITIONS OF THE CONTRACT*

00010	PRE-BID INFORMATION
00100	INSTRUCTIONS TO BIDDERS
00200	INFORMATION AVAILABLE TO BIDDERS
00300	BID FORMS
00400	SUPPLEMENTS TO BID FORMS
00500	AGREEMENT FORMS*
00600	BONDS AND CERTIFICATES*
00700	GENERAL CONDITIONS*
00800	SUPPLEMENTARY CONDITIONS*
00850	DRAWINGS AND SCHEDULES
00900	ADDENDA AND MODIFICATIONS

*These documents have important legal consequences. Initiation
or modification of them without explicit approval and guidance
of Owner or Owner's Counsel is not recommended.

SPECIFICATIONS-DIVISIONS 1-16

DIVISION 1 - GENERAL REQUIREMENTS

01010	SUMMARY OF WORK
01020	ALLOWANCES
01025	MEASUREMENT AND PAYMENT
01030	ALTERNATES/ALTERNATIVES
01040	COORDINATION
01050	FIELD ENGINEERING
01060	REGULATORY REQUIREMENTS
01070	ABBREVIATIONS AND SYMBOLS
01080	IDENTIFICATION SYSTEMS
01090	REFERENCE STANDARDS
01100	SPECIAL PROJECT PROCEDURES
01200	PROJECT MEETINGS
01300	SUBMITTALS
01400	QUALITY CONTROL
01500	CONSTRUCTION FACILITIES AND TEMPORARY CONTROLS
01600	MATERIAL AND EQUIPMENT
01650	STARTING OF SYSTEMS/COMMISSIONING
01700	CONTRACT CLOSEOUT
01800	MAINTENANCE

DIVISION 2 - SITE WORK

02010	SUBSURFACE INVESTIGATION
02050	DEMOLITION
02100	SITE PREPARATION
02140	DEWATERING
02150	SHORING AND UNDERPINNING
02160	EXCAVATION SUPPORT SYSTEMS
02170	COFFERDAMS
02200	EARTHWORK
02300	TUNNELING

02350	PILES AND CAISSONS
02450	RAILROAD WORK
02480	MARINE WORK
02500	PAVING AND SURFACING
02600	PIPED UTILITY MATERIALS
02660	WATER DISTRIBUTION
02680	FUEL DISTRIBUTION
02700	SEWERAGE AND DRAINAGE
02760	RESTORATION OF UNDERGROUND PIPELINES
02770	PONDS AND RESERVOIRS
02780	POWER AND COMMUNICATIONS
02800	SITE IMPROVEMENTS
02900	LANDSCAPING

DIVISION 3 - CONCRETE

03100	CONCRETE FORMWORK
03200	CONCRETE REINFORCEMENT
03250	CONCRETE ACCESSORIES
03300	CAST-IN-PLACE CONCRETE
03370	CONCRETE CURING
03400	PRECAST CONCRETE
03500	CEMENTITIOUS DECKS
03600	GROUT
03700	CONCRETE RESTORATION AND CLEANING
03800	MASS CONCRETE

DIVISION 4 - MASONRY

04100	MORTAR
04150	MASONRY ACCESSORIES
04200	UNIT MASONRY
04400	STONE
04500	MASONRY RESTORATION AND CLEANING
04550	REFRACTORIES
04600	CORROSION RESISTANT MASONRY

DIVISION 5 - METALS

05010	METAL MATERIALS
05030	METAL FINISHES
05050	METAL FASTENING
05100	STRUCTURAL METAL FRAMING
05200	METAL JOISTS
05300	METAL DECKING
05400	COLD-FORMED METAL FRAMING
05500	METAL FABRICATIONS
05580	SHEET METAL FABRICATIONS
05700	ORNAMENTAL METAL
05800	EXPANSION CONTROL
05900	HYDRAULIC STRUCTURES

DIVISION 6 - WOOD AND PLASTICS

06050	FASTENERS AND ADHESIVES
06100	ROUGH CARPENTRY
06130	HEAVY TIMBER CONSTRUCTION
06150	WOOD-METAL SYSTEMS
06170	PREFABRICATED STRUCTURAL WOOD
06200	FINISH CARPENTRY
06300	WOOD TREATMENT
06400	ARCHITECTURAL WOODWORK
06500	PREFABRICATED STRUCTURAL PLASTICS
06600	PLASTIC FABRICATIONS

DIVISION 7 - THERMAL AND MOISTURE PROTECTION

07100	WATERPROOFING
07150	DAMPPROOFING
07190	VAPOR AND AIR RETARDERS
07200	INSULATION
07250	FIREPROOFING
07300	SHINGLES AND ROOFING TILES
07400	PREFORMED ROOFING AND CLADDING/SIDING
07500	MEMBRANE ROOFING
07570	TRAFFIC TOPPING
07600	FLASHING AND SHEET METAL
07700	ROOF SPECIALTIES AND ACCESSORIES
07800	SKYLIGHTS
07900	JOINT SEALERS

DIVISION 8 - DOORS AND WINDOWS

08100	METAL DOORS AND FRAMES
08200	WOOD AND PLASTIC DOORS
08250	DOOR OPENING ASSEMBLIES
08300	SPECIAL DOORS
08400	ENTRANCES AND STOREFRONTS
08500	METAL WINDOWS
08600	WOOD AND PLASTIC WINDOWS
08650	SPECIAL WINDOWS
08700	HARDWARE
08800	GLAZING
08900	GLAZED CURTAIN WALLS

DIVISION 9 - FINISHES

09100	METAL SUPPORT SYSTEMS
09200	LATH AND PLASTER
09230	AGGREGATE COATINGS
09250	GYPSUM BOARD
09300	TILE
09400	TERRAZZO
09500	ACOUSTICAL TREATMENT
09540	SPECIAL SURFACES
09550	WOOD FLOORING
09600	STONE FLOORING

09630   UNIT MASONRY FLOORING
09650   RESILIENT FLOORING
09680   CARPET
09700   SPECIAL FLOORING
09780   FLOOR TREATMENT
09800   SPECIAL COATINGS
09900   PAINTING
09950   WALL COVERINGS

DIVISION 10 - SPECIALTIES

10100   CHALKBOARDS AND TACKBOARDS
10150   COMPARTMENTS AND CUBICLES
10200   LOUVERS AND VENTS
10240   GRILLES AND SCREENS
10250   SERVICE WALL SYSTEMS
10260   WALL AND CORNER GUARDS
10270   ACCESS FLOORING
10280   SPECIALTY MODULES
10290   PEST CONTROL
10300   FIREPLACES AND STOVES
10340   PREFABRICATED EXTERIOR SPECIALTIES
10350   FLAGPOLES
10400   IDENTIFYING DEVICES
10450   PEDESTRIAN CONTROL DEVICES
10500   LOCKERS
10520   FIRE PROTECTION SPECIALTIES
10530   PROTECTIVE COVERS
10550   POSTAL SPECIALTIES
10600   PARTITIONS
10650   OPERABLE PARTITIONS
10670   STORAGE SHELVING
10700   EXTERIOR SUN CONTROL DEVICES
10750   TELEPHONE SPECIALTIES
10800   TOILET AND BATH ACCESSORIES
10880   SCALES
10900   WARDROBE AND CLOSET SPECIALTIES

DIVISION 11 - EQUIPMENT

11010   MAINTENANCE EQUIPMENT
11020   SECURITY AND VAULT EQUIPMENT
11030   TELLER AND SERVICE EQUIPMENT
11040   ECCLESIASTICAL EQUIPMENT
11050   LIBRARY EQUIPMENT
11060   THEATER AND STAGE EQUIPMENT
11070   INSTRUMENTAL EQUIPMENT
11080   REGISTRATION EQUIPMENT
11090   CHECKROOM EQUIPMENT
11100   MERCANTILE EQUIPMENT
11110   COMMERCIAL LAUNDRY AND DRY CLEANING EQUIPMENT
11120   VENDING EQUIPMENT
11130   AUDIO-VISUAL EQUIPMENT

11140   SERVICE STATION EQUIPMENT
11150   PARKING CONTROL EQUIPMENT
11160   LOADING DOCK EQUIPMENT
11170   SOLID WASTE HANDLING EQUIPMENT
11190   DETENTION EQUIPMENT
11200   WATER SUPPLY AND TREATMENT EQUIPMENT
11280   HYDRAULIC GATES AND VALVES
11300   FLUID WASTE TREATMENT AND DISPOSAL EQUIPMENT
11400   FOOD SERVICE EQUIPMENT
11450   RESIDENTIAL EQUIPMENT
11460   UNIT KITCHENS
11470   DARKROOM EQUIPMENT
11480   ATHLETIC, RECREATIONAL AND THERAPEUTIC EQUIPMENT
11500   INDUSTRIAL AND PROCESS EQUIPMENT
11600   LABORATORY EQUIPMENT
11650   PLANETARIUM EQUIPMENT
11660   OBSERVATORY EQUIPMENT
11700   MEDICAL EQUIPMENT
11780   MORTUARY EQUIPMENT
11850   NAVIGATION EQUIPMENT

DIVISION 12 - FURNISHINGS

12050   FABRICS
12100   ARTWORK
12300   MANUFACTURED CASEWORK
12500   WINDOW TREATMENT
12600   FURNITURE AND ACCESSORIES
12670   RUGS AND MATS
12700   MULTIPLE SEATING
12800   INTERIOR PLANTS AND PLANTERS

DIVISION 13 - SPECIAL CONSTRUCTION

13010   AIR SUPPORTED STRUCTURES
13020   INTEGRATED ASSEMBLIES
13030   SPECIAL PURPOSE ROOMS
13080   SOUND, VIBRATION, AND SEISMIC CONTROL
13090   RADIATION PROTECTION
13100   NUCLEAR REACTORS
13120   PRE-ENGINEERED STRUCTURES
13150   POOLS
13160   ICE RINKS
13170   KENNELS AND ANIMAL SHELTERS
13180   SITE CONSTRUCTED INCINERATORS
13200   LIQUID AND GAS STORAGE TANKS
13220   FILTER UNDERDRAINS AND MEDIA
13230   DIGESTION TANK COVERS AND APPURTENANCES
13240   OXYGENATION SYSTEMS
13260   SLUDGE CONDITIONING SYSTEMS
13300   UTILITY CONTROL SYSTEMS
13400   INDUSTRIAL AND PROCESS CONTROL SYSTEMS
13500   RECORDING INSTRUMENTATION

13550   TRANSPORTATION CONTROL INSTRUMENTATION
13600   SOLAR ENERGY SYSTEMS
13700   WIND ENERGY SYSTEMS
13800   BUILDING AUTOMATION SYSTEMS
13900   FIRE SUPPRESSION AND SUPERVISORY SYSTEMS

DIVISION 14 - CONVEYING SYSTEMS

14100   DUMBWAITERS
14200   ELEVATORS
14300   MOVING STAIRS AND WALKS
14400   LIFTS
14500   MATERIAL HANDLING SYSTEMS
14600   HOISTS AND CRANES
14700   TURNTABLES
14800   SCAFFOLDING
14900   TRANSPORTATION SYSTEMS

DIVISION 15 - MECHANICAL

15050   BASIC MECHANICAL MATERIALS AND METHODS
15250   MECHANICAL INSULATION
15300   FIRE PROTECTION
15400   PLUMBING
15500   HEATING, VENTILATING, AND AIR CONDITIONING (HVAC)
15550   HEAT GENERATION
15650   REFRIGERATION
15750   HEAT TRANSFER
15850   AIR HANDLING
15880   AIR DISTRIBUTION
15950   CONTROLS
15990   TESTING, ADJUSTING, AND BALANCING

DIVISION 16 - ELECTRICAL

16050   BASIC ELECTRICAL MATERIALS AND METHODS
16200   POWER GENERATION
16300   HIGH VOLTAGE DISTRIBUTION (Above 600-Volt)
16400   SERVICE AND DISTRIBUTION (600-Volt and Below)
16500   LIGHTING
16600   SPECIAL SYSTEMS
16700   COMMUNICATIONS
16850   ELECTRIC RESISTANCE HEATING
16900   CONTROLS
16950   TESTING

Index

B

Backfill, 20
Board foot, 91–92
Borrow pit method, 14–17
Bricks:
    bonds, 55–56
    modular, 54, 59

C

Cabinets, 109
Carpet, 151
Cement, types of, 26
Checklists, 1, 4, 6
Construction Specifications Institute, 8
Convenience outlets, 166
Cost data, 4

D

Door types, 136–37
Drywall:
    fasteners, 146
    sizes, 147

E

Estimating process, 1–2
Excavation, 17

F

Fascia, 107
Final bid, 2, 175
Firestops, 98
Flashing, 131
Formwork, 26–28, 32
Frieze boards, 107

H

Heating and air conditioning, 157

I

Insulation, 127–28

J

Joints:
    construction, 43
    control, 44
Joists:
    ceiling, 98–99
    floor, 91

M

MASTERFORMAT, 8–9, 89, 145, 185–91

Metal decking, 83
Mortar, 66–67

N

National Electrical Code, 166

O

Overhead, 170–71

P

Painting, 148–49
Plumbing, 158
Plywood, 89, 93, 106–7
Profit, 1, 170–72
Proposal, 1, 3

R

Rafters, 100–3, 181–83
Reinforcement, 28–30, 65
Resilient flooring, 152
Roofing, 129–30

S

Siding, 107

Site clearing, 14
Slab types, 35–37
Soffits, 108–9
Specifications, 1–2
Stairs, 42–43, 109–10
Steel shapes, 80–81
Stretch-Out-Length, 9, 34, 176–79
Studs, 94, 96
Subcontractors:
    agreements, 5
    bids, 1, 2
Summary cost estimate, 1, 4, 7
Swellage and shrinkage, 18, 21
Switches, 166

T

Take-offs, 1, 4
Ties:
    form, 32–33
    masonry, 65

W

Water-cement ratio, 26
Waterproofing, 126–27
Window types, 141–42
Working drawings, 1–2